Lecture Notes in Mathematics 1797

Springer
Berlin
Heidelberg
New York
Hong Kong
London
Milan
Paris
Tokyo

Bernhard Schmidt

Characters and Cyclotomic Fields in Finite Geometry

Author

Bernhard Schmidt

Institut für Mathematik
Universität Augsburg
Universitätsstrasse 14
86135 Augsburg, Germany

E-mail: schmidt@math.uni-augsburg.de

Cataloging-in-Publication Data applied for.

Die Deutsche Bibliothek - CIP-Einheitsaufnahme

Schmidt, Bernhard:
Characters and cyclotomic fields in finite geometry / Bernhard Schmidt. -
Berlin ; Heidelberg ; New York ; Hong Kong ; London ; Milan ; Paris ; Tokyo
: Springer, 2002
(Lecture notes in mathematics ; 1797)
ISBN 3-540-44243-X

Mathematics Subject Classification (2000):
05B10, 05B20, 05B25

ISSN 0075-8434
ISBN 3-540-44243-x Springer-Verlag Berlin Heidelberg New York

Springer-Verlag Berlin Heidelberg New York a member of BertelsmannSpringer
Science + Business Media GmbH

http://www.springer.de

Typesetting: Camera-ready TEX output by the author

SPIN: 10891021 41/3142/du-543210 - Printed on acid-free paper

Acknowledgement

I would like to thank Siu Lun Ma and Clinton White for fruitful cooperation. Special thanks go to Dieter Jungnickel for help and encouragement.

Contents

Chapter 1

Introduction

This monograph contributes to the existence theory of combinatorial objects admitting certain types of automorphism groups. We will investigate (relative) difference sets, planar functions, group invariant weighing matrices, and two-weight irreducible cyclic codes. All these combinatorial objects have in common that they can be studied in terms of group ring equations, see Section 1.3. We use Fourier analysis on abelian groups to derive necessary conditions on the existence of the objects in question. This approach was already used in the fundamental work of Turyn [127] and dozens of subsequent papers on difference sets. However, all these results rely on severe technical conditions, the self-conjugacy assumption being the most infamous one, see Remark 1.4.4. The main merit of this monograph is the development of a method free from such severe assumptions and thus providing nonexistence theorems of broader applicability than all previously known results. We will obtain substantial progress towards three major conjectures which previously had seemed out of reach: The circulant Hadamard matrix conjecture, Ryser's conjecture and the Barker conjecture, see Sections 3.2.3, 3.2.4.

These results will be proved by the new method of the " field descent" which will be developed in Chapter 2. Roughly speaking, the field descent means that cyclotomic integers X for which $|X|^2$ is rational usually are contained in a much smaller cyclotomic field than a priori expected. The field descent not only can be used to prove nonexistence results for combinatorial objects, but also provides the probably most elementary approach to class group estimates for CM-fields. Our results on class number factors and bounds on p-ranks of class groups are comparable – in some cases even seem stronger – than those obtained by the usual methods of class field theory and Galois cohomology. Though our number theoretic results are only a by-product of our work, we believe that they establish an interesting connection between combinatorial and number theoretic questions.

Besides the field descent, this monograph contains two further major contri-

butions: An improvement of Turyn's self-conjugacy exponent bounds in Section 3.1 and a classification of two-weight irreducible cyclic codes in Chapter 4. The self-conjugacy exponent bounds are obtained by a substantial refinement of Turyn's method. In some cases, we obtain a dramatic improvement of Turyn's bounds and are able to provide necessary and sufficient conditions on the existence of certain difference sets. In Chapter 4, we use the Fourier analysis approach for the investigation of two-weight irreducible cylic codes. Though these objects have been studied in many papers (see [18]), a classification had not even been attempted yet. We will give a conjecturally complete classification and provide evidence for the completeness through theoretical results and computer searches.
The structure of this monograph is as follows. In Chapter 1, we introduce the combinatorial objects we will study and provide the necessary algebraic and number theoretic background. All results listed in Chapter 1 were previously known. In the following chapters, almost all results are new; previously known work is quoted as **"Result xyz"**. In Chapter 2, we develop the method of the field descent and give its applications to class group estimates. Chapter 3 contains the various exponent bounds we derive by Fourier analysis together with the field descent and algebraic-combinatorial methods. The results of Sections 3.1.2 3.1.3 and 3.1.4 are joint work with Siu Lun Ma. Chapter 4 on two-weight irreducible codes is joint work with Clinton White.

1.1 The nature of the problems

Good news! This section is readable for anyone who knows what a root of unity is. Through a hopefully well chosen example, we intend to give an impression of the nature of the combinatorial problems as well as of the typical methods the reader will be confronted with later. We will use only a minimum of terminology so that the material should be very easily accessible.
What we mainly will study is the *existence problem* for combinatorial structures invariant under certain operations, often called automorphisms. Such an existence problem usually is – if ever – decided in one of two ways: Positively through a construction of the desired object or negatively through a nonexistence proof. This monograph mainly contributes to the negative world though we also will have some petite positive news in the chapter on two-weight codes.
Let us illustrate the typical questions by an example. The combinatorial structure we consider is *Hadamard matrices*. An **Hadamard matrix of order v** is a $v \times v$-matrix H with entries ± 1 any two rows of which are orthogonal. The operation under which we require the Hadamard matrix to be invariant is something like cyclic shifting. More precisely, writing $H = (h_{i,j})_{i,j=0}^{v-1}$, we require $h_{i+1,j+1} = h_{i,j}$ for all i,j where the indices are taken modulo v. A matrix satisfying this condition is called **circulant**. It has the form

$$\begin{pmatrix} h_1 & h_2 & \cdots & h_v \\ h_v & h_1 & \cdots & h_{v-1} \\ \cdots & \cdots & \cdots & \cdots \\ h_2 & h_3 & \cdots & h_1 \end{pmatrix}$$

where we write $h_i := h_{0,i}$. There are trivial examples of circulant Hadamard matrices of orders 1 and 4:

$$(\ 1\ ), \quad \begin{pmatrix} -1 & 1 & 1 & 1 \\ 1 & -1 & 1 & 1 \\ 1 & 1 & -1 & 1 \\ 1 & 1 & 1 & -1 \end{pmatrix}.$$

Unfortunately, no circulant Hadamard matrix of any order $v \neq 1, 4$ has ever been found. This leads to the following conjecture first mentioned in Ryser's influential book [112, p. 134], but going back further to obscure sources.

Conjecture 1.1.1 (Circulant Hadamard matrix conjecture) *There is no circulant Hadamard matrix of order* $\neq 1, 4$.

Despite several claims of the contrary (see [79]), the circulant Hadamard conjecture has not been resolved yet. However, we will obtain substantial progress by using some algebraic tools which also work in a much broader context. Here we will describe how these algebraic tools come into play. We have chosen the example of circulant Hadamard matrices as it allows us to explain the algebraic approach very easily. We only use roots of unity, that's all what is needed to understand what is intended to be explained.
So, let $H = (h_{i,j})_{i,j=0}^{v-1}$ be a putative circulant Hadamard matrix of order v. Recall that $h_{i+1,j+1} = h_{i,j}$ and thus $h_{i+k,j+k} = h_{i,j}$ for all i, j, k by the definition of a circulant matrix. As usual, the indices are taken modulo v. We relabel the first row of H by $h_i := h_{0,i}$. Note $h_{i,j} = h_{j-i}$. Now, let ξ be any complex vth root of unity, define

$$X := \sum_{i=0}^{v-1} h_i \xi^i$$

and calculate the modulus of X:

$$\begin{aligned} |X|^2 &= \sum_{i=0}^{v-1} \left(\sum_{k=0}^{v-1} h_{i+k} h_k \right) \xi^i \\ &= \sum_{i=0}^{v-1} \left(\sum_{k=0}^{v-1} h_{-i,k} h_{0,k} \right) \xi^i \\ &= \sum_{i=0}^{v-1} (\delta_{i,0} v) \xi^i \\ &= v \end{aligned}$$

The third equality holds because of the orthogonality of the rows of H. This crucial necessary condition

$$X\overline{X} = v \tag{1.1}$$

(modulus equation) for the existence of a circulant Hadamard matrix can be viewed as an equation in the unknown X where $X \in \mathbb{Z}[\xi]$, the ring of algebraic integers of the cyclotomic field $\mathbb{Q}(\xi)$. The principal idea here is to use the fact that X generates a principal ideal $I := (X)$ satisfying $I\overline{I} = (v)$. Why is this fruitful? Because we have a complete knowledge of the ideal structure of $\mathbb{Z}[\xi]$ from algebraic number theory [13, 61, 109]. This can be used to get necessary conditions on the coefficients h_i of X. Actually, for many v, it is possible to show that at least one h_i must have modulus greater than 1 if (1.1) holds. This is the typical strategy for proving the nonexistence of circulant Hadamard matrices, see Chapter 3. In his important paper from 1965, Turyn [128] was the first to use this strategy which he calls the"size argument" [129]. For a long time it seemed that the last drop of the size argument had been milked — until it was revived in [117].

1.2 The combinatorial structures in question

In this section, we deal with the definitions and basic properties of the combinatorial structures which we will study.

1.2.1 Designs

Let m, n, k, λ be positive integers with $mn \geq k \geq 2$. A **divisible** $(\mathbf{m}, \mathbf{n}, \mathbf{k}, \lambda)$ **design** $\mathcal{D} = (\mathcal{P}, \mathcal{B})$ consists of a set $\mathcal{P}$ (points) together with a set $\mathcal{B}$ of subsets of $\mathcal{P}$ (blocks) such that the following conditions are satisfied.

- Each block consists of exactly k points.
- The point set $\mathcal{P}$ can be partitioned into m classes of size n such that any two points are contained in exactly λ common blocks if they are in different point classes and are contained in no common block if they are in the same point class.

A pair (p, B) with $B \in \mathcal{B}$ and $p \in B$ is called a **flag**. In the case $n = 1$, i.e., if each point class consists just of a single point, we write v instead of m and speak of a $(\mathbf{v}, \mathbf{k}, \lambda)$ **design**. In other words, for a (v, k, λ) design, the second condition above has to be replaced by "any two points are contained in exactly λ common blocks". The notion of a (v, k, λ) design is a natural generalization of the classical finite geometries constructed from finite fields, see Section 1.2.4. There is a vast literature on (v, k, λ) designs, see [12] for a recent, very readable and thorough treatment of most of the relevant results in this area. One of the central questions of design theory is "For what parameter triples (v, k, λ) does a (v, k, λ) design exist?". We begin with the easy part of the answer. By double counting all flags containing a given point and double counting all flags, one gets the following so-called **trivial necessary conditions**.

Lemma 1.2.1 *If a (v, k, λ) design exists, then*

$$\begin{aligned} \lambda(v-1) &\equiv 0 \pmod{k-1} \\ \lambda v(v-1) &\equiv 0 \pmod{k(k-1)}. \end{aligned}$$

Asymptotically in v, the extremely difficult existence problem for (v, k, λ) designs was completely solved by Wilson [131]. He developed a theory of "closed sets" which enabled him to show that the necessary conditions above are sufficient if v is large enough. More precisely:

Result 1.2.2 (Wilson) *For fixed k and λ, the necessary conditions for the existence of a (v, k, λ) design given in Lemma* 1.2.1 *are sufficient for all v greater than a constant only depending on k and λ.*

Though Result 1.2.2 is striking, many questions about (v, k, λ) designs remain open. This becomes clear, for instance, if we require additional properties for our designs. For example, we may ask for the number of points being the same as the number of blocks. Such a design is called **square** or — more commonly, but unfortunately — **symmetric**. Result 1.2.2 tells us nothing about symmetric (v, k, λ) designs. This is because the number of blocks of a (v, k, λ) design is $\lambda v(v-1)/k(k-1)$ which implies that for fixed k, λ and large v the design is not symmetric.
A very fruitful approach to symmetric designs is the use of there symmetries (or automorphisms) — though unfortunate symmetric designs may not have any symmetries.
We define automorphisms of general divisible designs now. An **automorphism** of a divisible design $\mathcal{D} = (\mathcal{P}, \mathcal{B})$ is a bijection $\tau : \mathcal{P} \cup \mathcal{B} \to \mathcal{P} \cup \mathcal{B}$ with $\tau(\mathcal{P}) = \mathcal{P}$ and $\tau(\mathcal{B}) = \mathcal{B}$ such that $p \in B$ if and only if $\tau(p) \in \tau(B)$ for all $p \in \mathcal{P}$ and all $B \in \mathcal{B}$. The set of all automorphisms of $\mathcal{D}$ forms a group Aut $\mathcal{D}$ called the **automorphism group** of $\mathcal{D}$. The group operation is composition. Any subgroup of Aut $\mathcal{D}$ is called **an automorphism group** of $\mathcal{D}$. An automorphism group G of $\mathcal{D}$ is called **point (block) regular** if for any two points $p, q \in \mathcal{P}$ (blocks $B, C \in \mathcal{B}$), there is exactly one $\tau \in G$ with $\tau(p) = q$ ($\tau(B) = C$). As a consequence of the so-called **orbit theorem**, see [12], we have the following.

Proposition 1.2.3 *An automorphism group of a symmetric (v, k, λ) design is point regular if and only if it is block regular.*

The strategy behind the use of automorphism groups for the construction of designs is the following. As an attempt to construct a (v, k, λ) design, one guesses a reasonable putative automorphism group G and only looks for designs having G as an automorphism group. Of course, compared to arbitrary (v, k, λ) designs, this gives valuable additional information making the search for the design easier. However, the major problem is if the choice of G was really clever, i.e., if a (v, k, λ) design with G as an automorphism group actually exists. This is one of the reasons why a classification of automorphism groups of designs is very desirable. There is an extensive literature on automorphism groups of symmetric designs. However, most problems still remain wide open. In Chapter 3, we will obtain some significant asymptotic results.

A crucial basic fact of the theory of designs is that a symmetric divisible design with a point regular automorphism group G can be completely described just by a certain subset of G corresponding to a block of the design. These subsets are the so-called **difference sets** which will be dealt with in the next section.

1.2.2 Difference Sets

Let G be a finite group of order nm, and let N be a subgroup of G of order n. A subset R of G is called an $(\mathbf{m}, \mathbf{n}, \mathbf{k}, \lambda)$ **difference set in G relative to N** if every $g \in G \setminus N$ has exactly λ representations $g = r_1 r_2^{-1}$ with $r_1, r_2 \in R$, and no nonidentity element of N has such a representation. The subgroup N is called the **forbidden subgroup**.

In the case $n = 1$, i.e., when the forbidden subgroup consists only of the identity element, we write v instead of m and speak of a $(\mathbf{v}, \mathbf{k}, \lambda)$ **difference set** in G. The nonnegative integer $n = k - \lambda$ is called the **order** of the difference set. If $n \in \{0, 1\}$, the difference set is called **trivial**. By a "difference set" we will always mean a nontrivial difference set. Sometimes it is useful to attach n to the parameters of a difference set. Thus we also speak of $(\mathbf{v}, \mathbf{k}, \lambda, \mathbf{n})$ **difference sets.**

Actually, difference sets, i.e., relative difference sets with $n = 1$, have been studied much more intensively than relative difference sets with $n > 1$. We refer the reader to [8, 12, 64, 65, 66, 70] for detailed treatments of the theory of difference sets. However, recently, the interest in relative difference sets has grown rapidly, see [6, 12, 23, 31, 32, 33, 35, 36, 48, 56, 75, 83, 84, 85, 87, 104, 105, 106, 107, 108, 111, 117].

The following well known result (see [12] or [105]) shows that relative difference sets are the same as symmetric divisible designs with point regular automorphism groups.

Proposition 1.2.4 *Let R be an (m, n, k, λ) difference set in G relative to N. Then $\mathcal{D} = (G, \{Rg : g \in G\})$ is a symmetric divisible (m, n, k, λ) design, and G is a point regular automorphism group of $\mathcal{D}$ acting by right translation. The point classes of $\mathcal{D}$ are the cosets of N.*

Conversely, let $\mathcal{D} = (\mathcal{P}, \mathcal{B})$ be a symmetric divisible (m, n, k, λ) design with a point regular automorphism group G. Fix a point $p \in \mathcal{P}$. For every $q \in \mathcal{P}$, let $g(q)$ be the unique element of G mapping p to q. Identify P and G via $q \leftrightarrow g(q)$. Let R be any block of $\mathcal{D}$, and let N be the point class containing the identity element of G. Then R is an (m, n, k, λ) difference set in G relative to N.

Because of its importance, we state the following special case of Proposition 1.2.4 separately.

Proposition 1.2.5 *A (v, k, λ) difference set is equivalent to a symmetric (v, k, λ) design with G as a point regular automorphism group.*

1.2.3 Projective planes and planar functions

The most beautiful and most thoroughly studied designs are projective planes. A **projective plane** is a symmetric (v, k, λ) design with $\lambda = 1$ which contains a quadrangle, i.e., four points no three of which are on a common block. It can be shown [12] that the parameters of a projective plane have the form $(v, k, \lambda) = (n^2 + n + 1, n + 1, 1)$ for some integer $n \geq 2$. The parameter n is called the **order** of the projective plane. By deleting one block (together with its points) of a projective plane of order n one obtains an **affine plane** of order n, i.e., an $(n^2, n, 1)$ design containing a triangle. This construction can be reversed. Thus an affine plane of order n exists if and only if there is a projective plane of order n. There is an extensive literature on projective planes and their automorphism groups. A standard reference is [58], see also [12, 37, 57, 105].
For any prime power q, there is an affine plane, and thus also a projective plane, of order n which can be constructed as follows. Let $\mathbb{F}_q$ be the finite field with q elements. Let $\mathcal{P} = \mathbb{F}_q^2$ be the two-dimensional vector space over $\mathbb{F}_q$, and let $\mathcal{B}$ be the set of all one-dimensional affine linear subspaces of $\mathbb{F}_q^2$. Then, by basic linear algebra, $\mathcal{D} = (\mathcal{P}, \mathcal{B})$ is an affine plane of order q.
On the other hand, not a single projective plane of non prime power order has been found yet. This leads us to the most important open problem in the theory of projective planes, the **prime power conjecture**.

Conjecture 1.2.6 *A projective plane of order n exists if and only if n is a prime power.*

What is known on the prime power conjecture? Basically, there are only three results:

1. There is an projective plane of order n for all prime powers n, see above.
2. Bruck-Ryser theorem [15]: If a projective plane of order $n \equiv 1$ or $2 \pmod 4$ exists, then n must be the sum of two squares.

3. There is no projective plane of order 10 (proved with massive help of a computer [69]).

The prime power conjecture far from being resolved, one can say more if one asks for projective planes with additional prescribed structure, for instance, certain automorphism groups. One problem of this type is the existence of **planar functions** which can be used for an elegant construction of affine planes with point regular automorphism groups. Planar functions are defined as follows. Let H and N be groups of order n. A mapping $f : H \to N$ is called a **planar function of degree n** if $h \mapsto f(gh)f(h)^{-1}$ is a bijection between H and N for every $g \in H \setminus \{1\}$. It is straightforward to check the following well known result (see [105]).

Proposition 1.2.7 *Let $f : H \to N$ be a planar function of degree n. Define*

$$\begin{aligned} \mathcal{P} : &= H \times N, \\ B(a,b) : &= \{(xa, f(x)b) : x \in H\}, \\ N(c) : &= \{(c,y) : y \in N\}, \\ \mathcal{B} : &= \{B(a,b) : a \in H, b \in B\} \cup \{N(c) : c \in H\}. \end{aligned}$$

Then $\mathcal{D} = (\mathcal{P}, \mathcal{B})$ is an affine plane of order n, and $H \times N$ is a point regular automorphism group of $\mathcal{D}$.

The standard example for a planar function is the mapping $f : (\mathbb{F}_q, +) \to (\mathbb{F}_q, +)$, $x \mapsto x^2$ where $(\mathbb{F}_q, +)$ is the additive group of the finite field with q elements, and q is an *odd* prime power. Here $x^2 := x \cdot x$ where "$\cdot$" is the multiplication in $\mathbb{F}_q$. To check that f is a planar function, we use the additive notation instead of the multiplicative notation in the definition of planar functions. Note that $h \mapsto f(g+h) - f(h) = (g+h)^2 - h^2 = 2gh + g^2$ is a nonconstant linear function and thus a bijection for all $g \neq 0$. Hence f indeed is a planar function.
As we explain now, planar functions are equivalent to certain relative difference sets and thus fit into the framework of our study. It is straightforward to check the following well known result (see [105]).

Lemma 1.2.8 *Let H and N be groups of order n. A mapping $f : H \to N$ is a planar function if and only if $R := \{(h, f(h)) : h \in H\}$ is an $(n,n,n,1)$ difference set in $H \times N$ relative to N.*

For *all known* planar functions, both H and N are elementary abelian p-groups with $p > 2$. Thus the following is conjectured.

Conjecture 1.2.9 *If a planar function $f : H \to N$ of degree n exists, then n is an odd prime power and both H and N are elementary abelian.*

Several partial results on Conjecture 1.2.9 can be found in the literature. It is easy to see that n has to be odd if a planar function $f : H \rightarrow N$ of degree n exists: Assume the contrary, and let R be the $(n, n, n, 1)$ difference set in $G := H \times N$ relative to N corresponding to f via Lemma 1.2.8. Let $g \in G \setminus N$ be an involution. Then there are $r_1, r_2 \in R$ with $g = r_1 r_2^{-1}$. However, this implies $g = g^{-1} = r_2 r_1^{-1}$ which means that g has at least two representation as a quotient of two elements of R. This is a contradiction to R being an $(n, n, n, 1)$ difference set in G relative to N. Thus no planar function of even degree can exist. For a generalization of this result, see [45, 63]. There are several further nonexistence results on planar functions, see [83, 105]. However, these results only apply to special cases of Conjecture 1.2.9. In particular, in cases where n has many distinct prime divisors, very little is known. In Chapter 3, we will obtain some significant progress by providing general exponent bounds on H and N. Compared to previously known results, these bounds are of broader applicability, especially for cases where n has many distinct prime divisors.

1.2.4 Projective geometries and Singer difference sets

Now we describe a classical construction of designs and difference sets which will play a role in our study of two-weight irreducible cyclic codes in Chapter 4. We begin with the construction of the classical finite projective spaces. Let q be a prime power. We denote the finite field with q elements by $\mathbb{F}_q$. Let $d \geq 2$ be an integer, and let $V = \mathbb{F}_q^{d+1}$ be the vector space of dimension $d+1$ over $\mathbb{F}_q$. Let $\mathcal{P}$ be the set of one-dimensional subspaces of V, and let $\mathcal{B}$ be the set of hyperplanes of V. Here we identify a hyperplane with the set of one-dimensional subspaces it contains. Then, by basic linear algebra, $\mathcal{D} = (\mathcal{P}, \mathcal{B})$ is a design with parameters

$$(v, k, \lambda) = \left(\frac{q^{d+1} - 1}{q - 1}, \frac{q^d - 1}{q - 1}, \frac{q^{d-1} - 1}{q - 1} \right). \tag{1.2}$$

This design is called the **classical point-hyperplane design**. The set of all nonzero subspaces of V is called the **Desarguesian projective geometry of dimension d over $\mathbb{F}_q$** and is denoted by $\mathrm{PG}(d, q)$.
As discovered by Singer [121] in 1938, the classical point-hyperplane designs have a cyclic, point regular automorphism group. This group is called the **Singer cycle** of $\mathrm{PG}(d, q)$. By Proposition 1.2.5, the existence of the Singer cycle implies the existence of a difference set with parameters (1.2) in a cyclic group. This difference set is called the **Singer difference set** of $\mathrm{PG}(d, q)$. Singer's result can be formulated elegantly in terms of the trace function $\mathrm{Tr} : \mathbb{F}_{q^{d+1}} \rightarrow \mathbb{F}_q$ as follows. We will give a proof as it is nice.

Theorem 1.2.10 *Let q be a prime power, let $d \geq 2$ be an integer. Then*

$$D := \{x\mathbb{F}_q^* : x \in \mathbb{F}_{q^{d+1}}^*, \mathrm{Tr}(x) = 0\}$$

is a difference set in $G := \mathbb{F}^*_{q^{d+1}}/\mathbb{F}^*_q$ *with parameters*

$$(v, k, \lambda) = \left(\frac{q^{d+1}-1}{q-1}, \frac{q^d-1}{q-1}, \frac{q^{d-1}-1}{q-1}\right).$$

Proof Since $\mathrm{Tr} : \mathbb{F}_{q^{d+1}} \to \mathbb{F}_q$ is surjective, we have $|\ker \mathrm{Tr}| = q^d$. Thus $|D| = (q^{d+1}-1)/(q-1)$ as required. Let $y \in \mathbb{F}^*_{q^{d+1}} \setminus \mathbb{F}^*_q$ be arbitrary. What we have to show is that there are exactly $\lambda(q-1)^2 = (q^{d-1}-1)(q-1)$ pairs (a, b) with $a, b \in \mathbb{F}^*_{q^{d+1}}$, $y\mathbb{F}^*_q = ab^{-1}\mathbb{F}^*_q$ and $a\mathbb{F}^*_q$, $b\mathbb{F}^*_q \in D$. These conditions are equivalent to $a, b \in \mathbb{F}^*_{q^{d+1}}$, $\mathrm{Tr}(yb) = 0$, $\mathrm{Tr}(b) = 0$ together with $a\mathbb{F}^*_q = yb\mathbb{F}^*_q$. Since $\{x \in \mathbb{F}^*_{q^{d+1}} : \mathrm{Tr}(yx) = 0\}$ and $\{x \in \mathbb{F}^*_{q^{d+1}} : \mathrm{Tr}(x) = 0\}$ are different hyperplanes of the $\mathbb{F}_q$-vector space $\mathbb{F}_{q^{d+1}}$ with intersection of cardinality q^{d-1}, there are exactly $q^{d-1} - 1$ solutions $b \in \mathbb{F}^*_{q^{d+1}}$ of $\mathrm{Tr}(yb) = 0$ and $\mathrm{Tr}(b) = 0$. For any such b, there are exactly $q - 1$ solutions $a \in \mathbb{F}^*_{q^{d+1}}$ of $a\mathbb{F}^*_q = yb\mathbb{F}^*_q$. Thus the number of pairs (a, b) satisfying the required conditions indeed is $(q^{d-1} - 1)(q - 1)$. □

1.2.5 Hadamard matrices and weighing matrices

Consider a $v \times v$ matrix H with real entries of absolute value at most 1. The absolute value of the determinant of H, i.e., the volume of the parallelopiped in the Euclidian v space spanned by the rows of H, is at most $v^{v/2}$ since each of these row vectors has length at most $\sqrt{v}$. This is a special case of Hadamard's well known bound [51] on determinants. Equality holds if and only if H has entries ± 1 only and any two rows of H are orthogonal. Such a matrix H is called a **Hadamard matrix** of order v. It satisfies the matrix equation $HH^t = vI$ where I is the identity matrix.
Let $(a_i), (b_i), (c_i)$ be three distinct rows of a Hadamard matrix H. Define $S := \sum_{i=1}^{v}(a_i + b_i)(a_i + c_i)$. Then $S \equiv 0 \pmod 4$ since $a_i + b_i, a_i + c_i \in \{-2, 0, 2\}$ for all i. On the other hand, by the orthogonaliy of the rows of H, we have $S = \sum_{i=1}^{v} a_i^2 = v$. This implies the following.

Lemma 1.2.11 *If a Hadamard matrix of order v exists, then $v = 1, 2$ or $v \equiv 0 \pmod 4$.*

There is a huge number of known construction methods for Hadamard matrices. It seems very likely that for all not too small $v \equiv 0 \pmod 4$, there are myriads of Hadamard matrices of order v.

Conjecture 1.2.12 *There is a Hadamard matrix of order v for all $v \equiv 0 \pmod 4$.*

Despite the multitude of constructions for Hadamard matrices, a proof a Conjecture 1.2.12 is still far out of sight. The density (in the integers) of the

orders v for which a Hadamard matrix of order v is known is 0. The smallest open case is $v = 428$.
As a generalization of Hadamard matrices, *weighing matrices* have been studied. Their properties are quite similar to those of Hadamard matrices, and they are important for some construction methods of Hadamard matrices, see [28]. A **weighing matrix** $\mathbf{W(m,n)}$ is an $m \times m$ matrix H with entries $-1, 0, 1$ only such that $HH^t = nI$ for some nonnegative integer n. We call n the **weight** of H. Weighing matrices have been investigated in many papers, see [47] for a survey, [40, 101, 102, 120] for some older and [27, 28, 50, 67, 73, 103] for some more recent results.
Now we turn our attention to weighing matrices which are invariant under a group operation. Let G be a group of order n. We say that an $m \times m$ matrix $H = (h_{f,g})_{f,g \in G}$ indexed with the elements of G is **G-invariant** if $h_{fk,gk} = h_{f,g}$ for all $f, g, k \in G$.
Let s_1 be the number of entries 1 in the first (and thus in any) row of a G-invariant weighing matrix $H = W(m,n)$. Let s be the sum of the entries of the first row of H (and thus of any row or column). Let I respectively J denote the identity matrix respectively the all-one matrix of order m. Then $(HH^t)J = H(H^tJ) = H(sJ) = s^2J$ and thus $n = s^2$ and $s = s_1 - (s^2 - s_1)$, i.e., $s_1 = s(s+1)/2$. So we have the following.

Lemma 1.2.13 *If a G-invariant weighing matrix $H = W(m,n)$ exists, then $n = s^2$ for some integer s, and the number of entries 1 in each row of H is $s(s+1)/2$.*

Note that we may assume that s is positive by replacing H by $-H$ if necessary. In Section 1.3, we will explain that group invariant Hadamard matrices are equivalent to certain difference sets called *Hadamard difference sets.*

1.2.6 Irreducible cyclic codes, two-intersection sets and sub-difference sets

In this section, we explain the equivalence of three notions which seem to be quite different at first sight: two-weight irreducible cyclic codes, subgroups of Singer cycles which are projective two-intersection sets and sub-difference sets of Singer difference sets. In Chapter 4, we will aim for a complete classification of these objects and will succeed conjecturally.
We begin with the coding theoretic formulation. For a standard reference on coding theory, see [80]. Let q be a prime power, and let $\mathbb{F}_q$ be the field with q elements. A **linear** $\mathbf{[n,k]}$ **code over** $\mathbb{F}_\mathbf{q}$ is a subspace C of the vector space $\mathbb{F}_q^n$ with $\dim C = k$. We make the standard assumption $(q,n) = 1$. The elements of C are called **codewords**. The **weight** of a codeword is the number of its nonzero entries. The **weight distribution** of C is the function $W : \mathbb{Z}_0^+ \to \mathbb{Z}_0^+$ where $W(i)$ is the number of codewords in C of weight i. If W takes at most two different nonzero values, we speak of a

two-weight code. A code C is called cyclic if $c = (c_0, ..., c_{n-1})^t \in C$ implies $(c_{n-1}, c_0, ..., c_{n-2})^t \in C$ for all $c \in C$. As usual in coding theory, we identify $\mathbb{F}_q^n$ with the residue class ring $\mathbb{F}_q[x]/(x^n - 1)$ via

$$a = (a_0, ..., a_{n-1})^t \leftrightarrow a(x) := \sum_{i=0}^{n-1} a_i x^i.$$

We have the following basic fact, see [80, Thm. 6.1.3].

Lemma 1.2.14 *A linear $[n, k]$ code over $\mathbb{F}_q$ is cyclic if and only if it is an ideal in $\mathbb{F}_q[x]/(x^n - 1)$.*

Let $C \subset \mathbb{F}_q[x]/(x^n - 1)$ be a cyclic code. By basic linear algebra [62], all elements of C are multiples of the monic polynomial in C of lowest degree. This polynomial g is called the **generator polynomial** of C. The dimension of C is $n - \deg g$, see [80, p. 77]. The generator polynomial divides $x^n - 1$ since otherwise the greatest common divisor of g and $x^n - 1$ would be a nonzero polynomial in C of lower degree than g.

Definition 1.2.15 *Let $C \subset \mathbb{F}_q[x]/(x^n - 1)$ be a cyclic code with generator polynomial g. If $g = (x^n - 1)/f$ where f is an irreducible divisor of $x^n - 1$, then C is called an* **irreducible cyclic code** *over $\mathbb{F}_q$ or a* **minimal cyclic code**.

We can restrict our attention to the *primitive* case, i.e., to irreducible cyclic codes for which the polynomial f is a minimal polynomial of a *primitive* nth root of unity over $\mathbb{F}_q$. To see this, assume that f is a minimal polynomial of a dth root of unity where d is any divisor of n. Then f divides $x^d - 1$ and thus $c(x)(x^d - 1) = 0$ in $\mathbb{F}_q[x]/(x^n - 1)$ for all codewords c. This means that the codewords of C are all periodic with period d, and the restriction of C to the first d components forms an irreducible cyclic code of lenght d with generator polynomial $(x^d - 1)/f$. This shows that "non-primitive" irreducible cyclic codes are just concatenations of "primitive" irreducible cyclic codes.
It is an interesting and in general extremely difficult problem to determine the weight distribution of irreducible cyclic codes. In Chapter 4, we will see that this problem is closely related to the evaluation of certain Gauss sums. For the time being, we give a useful alternative description of irreducible cyclic codes using the trace function. For the convenience of the reader, we include a proof for the correctness. As explained above, we only consider the case where the polynomial f is a minimal polynomial of a primitive nth root of unity.

Proposition 1.2.16 *Let q be a prime power, and let n be a positive integer with $(n, q) = 1$. Write $k := \mathrm{ord}_n(q)$, and let β be a primitive nth root of unity in $\mathbb{F}_{q^k}$. Let Tr denote the trace function from $\mathbb{F}_{q^k}$ to $\mathbb{F}_q$. The set*

$$C_\beta := \{c(y) := (\mathrm{Tr}(y), \mathrm{Tr}(y\beta), ..., \mathrm{Tr}(y\beta^{n-1}))^t : y \in \mathbb{F}_q^k\} \tag{1.3}$$

is an irreducible cyclic $[n, k]$ *code over* $\mathbb{F}_q$. *Moreover, any irreducible cyclic code can be obtained in this way.*

Proof We first show that C_β is an irreducible cyclic $[n, k]$ code. The linearity of C_β follows from the linearity of the trace function. Since $c(y\beta^{-1})$ is a cyclic shift of $c(y)$, the code C_β is cyclic. As $\mathbb{F}_{q^k}$ is the splitting field of $x^n - 1$ over $\mathbb{F}_q$, we know that β is a root of an irreducible polynomial h over $\mathbb{F}_q$ of degree k. Note that $f := h(x^{-1})x^k$ is the minimal polynomial of β^{-1} and thus is irreducible. A quick calculation using the linearity of the trace function shows that $c(x)f(x) = 0$ in $\mathbb{F}_q[x]/(x^n - 1)$ for all $c \in C_\beta$. Thus $(x^n - 1)/f$ is the generator polynomial for C_β. The dimension of C is k since deg $f = k$.
Now let C be an arbitrary irreducible cyclic $[n, k]$ code over $\mathbb{F}_q$ where $(n, q) = 1$. Let $(x^n - 1)/f$ be the generator polynomial for C where f is the minimal polynomial of a primitive nth root α over $\mathbb{F}_q$. Then $k = \dim C = \deg f = \mathrm{ord}_n(q)$, the last equality being a basic result on finite fields, see [76]. Using the first part of the proof, we get

$$C = \{c(y) \in \mathbb{F}_q[x]/(x^n - 1) : c(y)f(y) = 0\} = C_{\alpha^{-1}}.$$

□

Proposition 1.2.16 shows that we can restrict our study of irreducible cyclic codes to the case where n is divisible by $q - 1$, that is, the case where $\langle\beta\rangle$ contains $\mathbb{F}_q^*$. To see this, let γ be a generator of $\langle\beta\rangle\mathbb{F}_q^*$ and observe that, by the linearity of the trace function, the codewords of C_γ can be obtained by multiplying codewords of C_β with suitable elements of $\mathbb{F}_q^*$ and linking the results in the right way. Thus, from now on, we assume that n is divisible by $q - 1$.
Next we explain the equivalence between irreducible cyclic two-weight codes and subgroups of Singer cycles which are projective two-intersection sets. A **projective** $(\mathbf{h}, \mathbf{d}, \mathbf{h_1}, \mathbf{h_2})$ **set** over $\mathbb{F}_q$ is a proper, nonempty set H of h points of $\mathrm{PG}(d - 1, q)$ such that every hyperplane of $\mathrm{PG}(d - 1, q)$ meets H in h_1 or h_2 points. We will call such a set H a **projective two-intersection set.**
Sometimes, H is also called a **set of type** $(\mathbf{h_1}, \mathbf{h_2})$. We refer the reader to [18] for a thorough treatment of projective two-intersection sets and their connections to two-weight codes and strongly regular graphs. Projective two-intersection sets which are unions of cosets of a subgroup of the Singer cycle of $\mathrm{PG}(d - 1, q)$ are called **cyclotomic**, see [18, Section 9].

Proposition 1.2.17 *The code* C_β *given by* (1.3) *is a two-weight code if and only if* $\langle\beta\rangle$ *is a projective two-intersection set in* $PG(k - 1, q)$.

Proof The weight of the codeword $c(y)$ of C_β is $n - (q - 1)t(y)$ where $t(y)$ is the number of points in $\langle\beta\rangle$ which are contained in the hyperplane

$H(y) = \{z \in \mathbb{F}_{q^k} : \mathrm{Tr}(zy) = 0\}$. Note that $H(y)$ runs over all hyperplanes of $\mathrm{PG}(k-1, q)$ when y runs over $\mathbb{F}_{q^k}$. Thus C_β is a two-weight code if and only if $\langle\beta\rangle$ has at most two different intersection numbers with the hyperplanes of $\mathrm{PG}(k-1, q)$. □

Now we come to the connection to difference sets. The following observation is basically due to McFarland [97]. We will give a proof in Section 1.3.

Lemma 1.2.18 *Let D be a difference set in a group G, and let N be a normal subgroup of G. If D has at most two different intersection numbers a, b with the cosets of N in G, then*

$$\overline{D} := \{Ng : g \in G, |D \cap Ng| = a\}$$

is a difference set in G/N.

The difference set $\overline{D}$ appearing in Lemma 1.2.18 is called a **sub-difference set** of D in G/N, see [97]. We conclude this section by showing that sub-difference sets of Singer difference sets and two-weight irreducible cyclic codes are equivalent objects.

Proposition 1.2.19 *The code C_β given by (1.3) is a two-weight code if and only if the Singer difference set*

$$D = \{d\mathbb{F}_q^* : d \in \mathbb{F}_{q^k}^*, \mathrm{Tr}(d) = 0\}$$

of $PG(k-1, q)$ has a sub-difference set in $G/\langle\beta\rangle$ where $G = \mathbb{F}_{q^k}^*/\mathbb{F}_q^*$ is the Singer cycle of $PG(k-1, q)$.

Proof The set of hyperplanes of $\mathrm{PG}(k-1, q)$ is

$$\{Dg : g \in \mathbb{F}_{q^k}^*\},$$

see Sections 1.2.1 and 1.2.2. The intersection numbers of D with the cosets of $\langle\beta\rangle$ coincide with the intersection numbers of $\langle\beta\rangle$ with the hyperplanes of $\mathrm{PG}(k-1, q)$:

$$|D \cap \langle\beta\rangle g| = |Dg^{-1} \cap \langle\beta\rangle|.$$

Thus the assertion follows from Proposition 1.2.17 and the definition of a sub-difference set. □

1.3 Group rings, characters, Fourier analysis

Consider a finite group G of order v acting regularly on a set $\mathcal{P}$ of v objects, for instance, the points of a design. The regularity of G allows us to identify

$\mathcal{P}$ with G: Fix a point $p \in \mathcal{P}$ and identify an arbitrary point q with the unique element $g_q \in G$ mapping p to q. Under this identification, the blocks of the design become (relative) difference sets in G, see Section 1.2.2. Difference sets in G can be characterized by an equation over the integral group ring $\mathbb{Z}[G]$. This means that point regular automorphism groups of designs are equivalent to certain solutions of group ring equations. Though this essentially only is a change of notation, the use of group rings is almost indispensable because of its conciseness and elegance.

To explain the group ring language, we first fix some notation. We will always identify a subset A of a group G with the element $\sum_{g \in A} g$ of $\mathbb{Z}[G]$. Let e_G be the identity element of G. For $k \in \mathbb{Z}$ we simply write k instead of $ke_G \in \mathbb{Z}[G]$. For $B = \sum_{g \in G} b_g g \in \mathbb{Z}[G]$ define $B^{(-1)} := \sum_{g \in G} b_g g^{-1}$ and $|B| := \sum_{g \in G} b_g$. A group homomorphism $G \to H$ is always assumed to be extended to a homomorphism $\mathbb{Z}[G] \to \mathbb{Z}[H]$ by linearity. Now we are ready to formulate the difference set property as a group ring equation.

Lemma 1.3.1 *A k-subset R of a group G of order mn is a (m, n, k, λ) difference set in G relative to a subgroup N if and only if*

$$RR^{(-1)} = k + \lambda(G - N)$$

in $\mathbb{Z}[G]$.

Proof If we expand $RR^{(-1)}$, we get the sum S over all quotients $r_1 r_2^{-1}$, $r_1, r_2 \in R$, $r_1 \neq r_2$, plus k times the identity element. By definition, the set R is a (m, n, k, λ) difference set in G relative to a N if and only if $S = \lambda(G - N)$ in $\mathbb{Z}[G]$. □

We state the difference set case of Lemma 1.3.1 separately.

Lemma 1.3.2 *A k-subset D of a group G of order v is a (v, k, λ, n) difference set in G if and only if*

$$DD^{(-1)} = n + \lambda G$$

in $\mathbb{Z}[G]$.

A very fruitful approach to the study of difference sets is to apply group representations φ to the group ring equation in Lemma 1.3.1. The first crucial step of this algebraic approach is to get necessary conditions on the images $\varphi(R)$ of the difference set by using some machinery from algebra and number theory. In this context, several basic questions are still wide open. In some situations, it is even not known how to get *any useful information* via algebra or number theory. However, we feel that it should be possible to broaden the algebraic approach, even beyond the new subfield strategy we will present in Section 2.

The second crucial step of the algebraic approach is to retranslate the necessary conditions into information on D itself. This often includes finding a

way to combine information on D one gets from different representations φ. This usually is — if at all — hard to achieve. In Chapter 3 we will see some nice examples of a successful application of this method.
In the present monograph, only 1-dimensional complex group representations, i.e., complex characters, will be used. This still will allow us to obtain some significant nonexistence results on (relative) difference sets in nonabelian groups. However, for a more thorough study of difference sets in nonabelian groups, especially for *construction purposes*, non-linear representations are indispensable, see [59, 60, 77, 78].
Before we explain the use of characters in the study of (relative) difference sets we review some important results on characters and Fourier transforms. For the basic facts on characters of finite abelian groups see [90], for instance. All groups we consider in the following are finite. For an abelian group G, we denote its character group by G^*, and for a subgroup U of G, we write $U^\perp$ for the subgroup of all characters which are trivial on U. For a subgroup W of G^*, we write

$$W^\perp := \{g \in G : \chi(g) = 1 \text{ for all } \chi \in W\}.$$

The following **orthogonality relations** are extremely useful.

Lemma 1.3.3 *Let G be an abelian group, let U be a subgroup of G, and let W be a subgroup of G^*. Then*
a) $\sum_{g\in U} \chi(g) = 0$ for all $\chi \in G^* \setminus U^\perp$ and
b) $\sum_{\chi\in W} \chi(g) = 0$ for all $g \in G \setminus W^\perp$.

As a consequence of the orthogonality relations, one gets the so-called **Fourier inversion formula**.

Lemma 1.3.4 *Let G be an abelian group, and let $A = \sum_{g\in G} a_g g \in \mathbb{Z}[G]$. Then*

$$a_g = \frac{1}{|G|} \sum_{\chi\in G^*} \chi(Ag^{-1})$$

for all $g \in G$.

Corollary 1.3.5 *Let G be an abelian group. If $A \in \mathbb{Z}[G]$ with $\chi(A) = 0$ for all nontrivial characters χ of G then $A = \mu G$ for some integer μ.*

The Fourier inversion formula leads to the following criterion for a subset of an abelian group to be a relative difference set.

Lemma 1.3.6 *A k-subset R of an abelian group G of order mn is an (m, n, k, λ)-difference set in G relative to a subgroup N of order n if and only if*

$$\chi(R)\overline{\chi(R)} = \begin{cases} k & \text{if } \chi \in G^* \setminus N^\perp \\ k - \lambda n & \text{if } \chi \in N^\perp \end{cases} \tag{1.4}$$

for every nontrivial character χ of G.

Proof By Lemma 1.3.1, we have to show that (1.4) is equivalent to

$$RR^{(-1)} = k + \lambda(G - N). \tag{1.5}$$

Using the orthogonality relations 1.3.3, we see that (1.5) implies (1.4). Now assume that (1.4) holds, i.e., the character values of both sides of (1.5) are the same except possibly for the trivial character. Corollary 1.3.5 implies $RR^{(-1)} = k + \lambda(G - N) + \mu G$ for some integer μ. Comparing the coefficient of 1 in this equation, we see that $\mu = 0$, i.e., (1.5) holds. □

As usual, we state the difference set case separately.

Lemma 1.3.7 *A k-subset D of an abelian group of order v is a (v, k, λ, n) difference set in G if and only if*

$$\chi(D)\overline{\chi(D)} = n$$

for every nontrivial character χ of G.

If we are only looking for *necessary* conditions for the existence of (relative) difference sets, we can allow nonabelian groups and consider characters of abelian homomorphic images. We get the following result whose proof is essentially contained in the proof of Lemma 1.3.6.

Lemma 1.3.8 *Let G be a group of order nm, let N be a subgroup of G of order n, and let R be an (m, n, k, λ)-difference set in G relative to N. Let U be a normal subgroup of G such that G/U is abelian and let $\rho : G \to G/U$ be the canonical epimorphism. Then*

$$\chi(\rho(R))\overline{\chi(\rho(R))} = \begin{cases} k & \text{if } \chi \in (G/U)^* \setminus N_U^\perp \\ k - \lambda n & \text{if } \chi \in N_U^\perp \end{cases}$$

for every nontrivial character χ of G/U where $N_U = \{Un : n \in N\}$.

Group invariant weighing matrices (see Section 1.2.5) also allow a concise description in terms of group rings. To this end, we identify a G-invariant matrix $H = (h_{g,k})$ with the element $\sum_{g \in G} h_{1,g} g$ of $\mathbb{Z}[G]$.

Lemma 1.3.9 *An element H of $\mathbb{Z}[G]$ with coefficients $-1, 0, 1$ is a G-invariant weighing matrix $W(m, n)$ if and only if*

$$HH^{(-1)} = n \tag{1.6}$$

in $\mathbb{Z}[G]$.

Proof The equation $HH^t = nI$ is equivalent to

$$\sum_{g\in G} h_{i,g}h_{j,g} = \sum_{g\in G} h_{1,i^{-1}g}h_{1,j^{-1}g} = \delta_{ij}n$$

for all $i,j \in G$ where δ_{ij} is the Kronecker symbol. Thus H is a G-invariant weighing matrix $W(m,n)$ if and only if $HH^{(-1)} = \sum_{g,k\in G} h_{1,g}h_{1,k}gk^{-1} = \sum_{l\in G}\left(\sum_{g\in G} h_{1,g}h_{1,l^{-1}g}\right)l = n$ in $\mathbb{Z}[G]$. □

Corollary 1.3.10 *Assume that a weighing matrix $H = W(m,n)$ is G-invariant. Let U be a subgroup of G such that G/U is abelian. Let $\rho : G \to G/U$ denote the canonical epimorphism. Then*

$$\chi(\rho(H))\overline{\chi(\rho(H))} = n$$

for all characters χ of G/U where H is viewed as an element of $\mathbb{Z}[G]$.

Proof This follows by applying $\chi \circ \rho$ to (1.6). □

A **Hadamard difference set** is a difference set with parameters $(v,k,\lambda,n) = (4u^2, 2u^2-u, u^2-u, u^2)$ for some integer u. Lemmas 1.3.2 and 1.3.9 enable us to show that Hadamard difference sets and group invariant Hadamard matrices are equivalent objects.

Lemma 1.3.11 *A G-invariant Hadamard matrix of order $|G| > 1$ exists if and only if there is an Hadamard difference set in G. In particular, the order of a group invariant Hadamard matrix is an even square.*

Proof First assume that H is a G-invariant Hadamard matrix. By Lemmas 1.2.11 and 1.2.13, $|G|$ is an even square, say $|G| = 4u^2$ for some u. View H as an element of $\mathbb{Z}[G]$ as in Lemma 1.3.9. Define $D := (H+G)/2$. Since H has coefficients ± 1 only, D is an element of $\mathbb{Z}[G]$ with coefficients $0,1$ only, i.e., D is a subset of G. By Lemma 1.2.13, replacing H by $-H$ if necessary, the number of elements of D, i.e., the number of entries 1 in the first row of H, is $2u^2-u$. Note $|H| = -2u$ when H is viewed as a group ring element. Using Lemma 1.3.9, we get

$$\begin{aligned} 4DD^{(-1)} &= HH^{(-1)} + (4u^2-4u)G \\ &= 4u^2 + (4u^2-4u)G. \end{aligned}$$

Thus D is a Hadamard difference set in G by Lemma 1.3.2. Conversely, assume that D is a Hadamard difference set in G. Then, by Lemmas 1.3.2 and 1.3.9, $H := 2D - G$ corresponds to a G-invariant Hadamard matrix. □

Concluding this group ring section, we give a proof for the sub-difference set Lemma 1.2.18.

Proof of Lemma 1.2.18 Let D be a (v, k, λ, n) difference set in a group G, and let N be a normal subgroup of G such that D has only two different intersection numbers a, b with the cosets of N. Let $\rho : G \to G/N$ be the canonical epimorphism. W.l.o.g assume $a > b$ and write $\rho(D) = (a-b)E + b(G/N)$ where

$$E := \{Ng \in G/N : |D \cap Ng| = a\}.$$

From Lemma 1.3.2 we have $DD^{(-1)} = n + \lambda G$. Applying ρ to this equation, we get

$$\begin{aligned} n + \lambda|N|(G/N) &= \rho(D)\rho(D)^{(-1)} \\ &= [2(ab - b^2)|E| + b^2|G/N|](G/N) \\ &\quad +(a-b)^2 EE^{(-1)}. \end{aligned}$$

Thus

$$EE^{(-1)} = n' + \lambda'(G/N)$$

for some nonnegative integers λ', n'. Lemma 1.3.2 shows that E indeed is a difference set in G/N. □

1.4 Number theoretic tools

The main part of this section deals with cyclotomic integers of prescribed absolute value, i.e., solutions $X \in \mathbb{Z}[\xi]$ of an equation $X\overline{X} = n$, $n \in \mathbb{Z}^+$, where ξ is a complex root of unity. We will list some well known necessary conditions on the solutions X. We also recall some facts on the most famous examples of cyclotomic integers of prescribed absolute value, namely, *Gauss sums.* Gauss sums will be used in our study of irreducible cyclic two-weight codes, see Chapter 4.

As we have explained in Section 1.3, the algebraic approach to (relative) difference sets consists of two main steps. Here we concentrate on the first step, i.e., obtaining information on the image of a putative difference set D under group representations χ. We will only consider the case where χ is a complex character. Thus, by Lemma 1.3.6, we have to study equations of the form

$$X\overline{X} = n \tag{1.7}$$

where n is a positive integer and $X \in \mathbb{Z}[\xi_m]$ for some m, $\xi_m = \exp 2\pi i/m$. We call (1.7) the **modulus equation**.

What is known about the modulus equation? Not much! We summarize the most important facts here. In Chapter 2, we will present new results obtained by what we call the "field descent". We will also show that the modulus equation (1.7) is closely related to the class group of $\mathbb{Q}(\xi_m)$. The classical result on the modulus equation is the following lemma of Kronecker's. See [13, Section 2.3, Thm. 2] for a proof.

Lemma 1.4.1 *An algebraic integer all of whose conjugates have absolute value 1 is a root of unity.*

Note that Lemma 1.4.1 implies that any cyclotomic integer of absolute value 1 must be a root of unity since the Galois group of a cyclotomic field is abelian.
To state the further results on the modulus equation, we need some notation. By $\mathbb{Q}(\xi_m)$, $\xi_m = e^{2\pi i/m}$, we denote the mth cyclotomic field over $\mathbb{Q}$. By a fundamental result of algebraic number theory [109, p. 269, Thm. 4B (3)] the ring of algebraic integers of $\mathbb{Q}(\xi_m)$ is $\mathbb{Z}[\xi_m]$. For the basic properties of $\mathbb{Z}[\xi_m]$, see [61, Chapter 12], for instance. For $\sigma \in \mathrm{Gal}(\mathbb{Q}(\xi_m)/\mathbb{Q})$, we write $\mathrm{Fix}(\sigma)$ for the subfield of $\mathbb{Q}(\xi_m)$ fixed by $\langle\sigma\rangle$. For relatively prime integers t and s, we denote the multiplicative order of t modulo s by $\mathrm{ord}_s(t)$. Finally, φ denotes the Euler totient function.
The following standard result completely determines the factorization pattern of rational primes in cyclotomic fields. See [89, Thm. 8.8] for a proof.

Lemma 1.4.2 *Let m be a positive integer, and let p be a prime. Write $m = p^a m'$ with $(m', p) = 1$ and $a \geq 0$. Then p factors in $\mathbb{Q}(\xi_m)$ as*

$$(p) = \prod_{i=1}^{t} \pi_i^{\varphi(p^a)}$$

where $t = \varphi(m')/\mathrm{ord}_{m'}(p)$, and the π_i are distinct prime ideals.

We recall that the **decomposition group** of a prime ideal P of $\mathbb{Z}[\xi_m]$ is the set of all $\sigma \in \mathrm{Gal}(\mathbb{Q}(\xi_m)/\mathbb{Q})$ with $P^\sigma = P$. It has been known for half a century that decomposition groups play a central role in the study of difference sets. The most important contributions in this context are due to Hall [52] (multiplier theorem), Yamamoto [133] (exploitation of extremely large decomposition groups) and Turyn [127] (self-conjugacy condition). In Chapters 2 and 3, we will be able to make use of parts of decomposition groups which have not been exploited yet.
Many results of this monograph rely on the following complete description of the decomposition groups of prime ideals of cyclotomic fields. For the convenience of the reader, we include a proof.

Theorem 1.4.3 *Let p be a rational prime, let P be a prime ideal above p in $\mathbb{Z}[\xi_m]$, and write $m = p^a m'$ with $(m', p) = 1$. The decomposition group of P consists of all $\sigma \in \mathrm{Gal}(\mathbb{Q}(\xi_m)/\mathbb{Q})$ for which there is an integer j such that*

$$\sigma(\xi_{m'}) = \xi_{m'}^{p^j}. \tag{1.8}$$

Proof The ideal $(1-\xi_{p^a})$ of $\mathbb{Z}[\xi_m]$ is divisible by P since $(p) = (1-\xi_{p^a})^{(p-1)p^{a-1}}$ if $a > 0$, see [89, (8.24)] or [12, XI.§13]. Hence $(\xi_{p^a}^i)^\tau \equiv 1 \pmod{P}$ for all $i \in \mathbb{Z}$, and all $\tau \in \mathrm{Gal}(\mathbb{Q}(\xi_m)/\mathbb{Q})$. Let A be any element of $\mathbb{Z}[\xi_m]$, and write $A = \sum_{i=0}^{p^a-1} \xi_{p^a}^i f_i(\xi_{m'})$ with $f_i \in \mathbb{Z}[x]$. If $\sigma \in \mathrm{Gal}(\mathbb{Q}(\xi_m)/\mathbb{Q})$ satisfies (1.8) then

$$\begin{aligned}
A^\sigma &= \sum_{i=0}^{p^a-1} (\xi_{p^a}^i)^\sigma f_i(\xi_{m'}^{p^j}) \\
&\equiv \sum_{i=0}^{p^a-1} f_i(\xi_{m'}^{p^j}) \\
&\equiv (\sum_{i=0}^{p^a-1} f_i(\xi_{m'}))^{p^j} \\
&\equiv (\sum_{i=0}^{p^a-1} \xi_{p^a}^i f_i(\xi_{m'}))^{p^j} \\
&\equiv A^{p^j} \pmod{P}.
\end{aligned}$$

Now, $A \in P$ implies $A^{p^j} \in P$ and thus $A^\sigma \in P$. Hence $P^\sigma \subset P$, implying $P^\sigma = P$ since P^σ is a prime ideal and thus maximal [61, p. 177, Cor. 2]. Thus σ fixes P if it satisfies (1.8). Note that the number of $\sigma \in \mathrm{Gal}(\mathbb{Q}(\xi_m)/\mathbb{Q})$ satisfying (1.8) is $\varphi(p^a)\mathrm{ord}_{m'}(p)$. By the orbit formula, there cannot be any further Galois automorphism of $\mathbb{Q}(\xi_m)$ fixing P since $\mathrm{Gal}(\mathbb{Q}(\xi_m)/\mathbb{Q})$ acts transitively on the set of prime ideals dividing p [61, Prop. 12.3.3] and since there are exactly $\varphi(m')/\mathrm{ord}_{m'}(p)$ of these ideals, see Lemma 1.4.2. □

Remark 1.4.4 By Theorem 1.4.3 a prime ideal P above p in $\mathbb{Q}(\xi_m)$ is invariant under complex conjugation if and only if there is an integer j with $p^j \equiv -1 \pmod{m'}$. In this case p is called **self-conjugate modulo** m, see [65, 105, 127]. A composite integer n is called self-conjugate modulo m if every prime divisor of n has this property. One can see that a prime p can only be self-conjugate modulo m if the exact power of 2 dividing $\mathrm{ord}_q(p)$ is the same for all prime divisors $q \neq p$ of m. Thus, loosely speaking, the probability that n is self-conjugate modulo m decreases exponentially fast in the number of distinct prime divisors of n and in the number of distinct prime divisors of m.

As a consequence of Theorem 1.4.3, we get the following result whose importance for the study of difference sets first was noticed by Turyn [127].

Corollary 1.4.5 *Assume that* $A \in \mathbb{Z}[\xi_m]$ *satisfies*

$$A\overline{A} \equiv 0 \bmod t^{2b}$$

where b, t are positive integers, and t is self-conjugate modulo m. Then

$$A \equiv 0 \bmod t^b.$$

Proof By Theorem 1.4.3 the prime ideals above t in $\mathbb{Z}[\xi_m]$ are invariant under complex conjugation. □

Now we recall some facts on Gauss sums which will be needed in Chapter 4. We will use the notation $\xi_t := \exp 2\pi i/t$.

Definition 1.4.6 *Let $r = p^a$ be a prime power, and let χ be a character of $\mathbb{F}_r^*$. We define*

$$G(\chi) := \sum_{x \in \mathbb{F}_r^*} \chi(x) \xi_p^{\mathrm{Tr}(x)}$$

where Tr *denotes the (absolute) trace from $\mathbb{F}_r$ to $\mathbb{F}_p$.*

Through the following Lemma we see that Gauss sums are solutions of the modulus equation (1.7). A proof can be found in [76, Thm. 5.11], for instance.

Lemma 1.4.7 *If χ is nontrivial, then*

$$|G(\chi)|^2 = r.$$

It was shown in [132] that, up to a constant factor, the character values of Singer difference sets are Gauss sums. Because of its importance for Chapter 4, we include a proof of this result here.

Lemma 1.4.8 *Let q be a prime power, and let $k \geq 3$ be an integer. Let χ be a character of $\mathbb{F}_{q^k}^*$ of order d dividing $(q^k - 1)/(q - 1)$. Note that χ is trivial on $\mathbb{F}_q^*$ and thus can be viewed as a character of the Singer cycle $G = \mathbb{F}_{q^k}^*/\mathbb{F}_q^*$ of* PG$(k-1, q)$. *Let D be the Singer difference set in G corresponding to* PG$(k-1, q)$. *Then*

$$G(\chi) = q\chi(D).$$

In particular, $G(\chi) \in \mathbb{Z}[\xi_d]$.

Proof Write $E := \mathbb{F}_{q^k}$ and $F := \mathbb{F}_q$. We recall that

$$D = \{x\mathbb{F}^* : x \in E^*, \mathrm{Tr}_{E/F}(x) = 0\}.$$

Using the triviality of χ on F^*, the linearity and transitivity of the trace, the fact that $\sum_{x\in F^*}\xi_p^{\mathrm{Tr}(x)} = -1$, and the orthogonality relations, we calculate

$$\begin{aligned}
G(\chi) &= \sum_{x\in E^*}\chi(x)\xi_p^{\mathrm{Tr}(x)}\\
&= \sum_{a\in E^*/F^*}\chi(a)\sum_{b\in F^*}\xi_p^{\mathrm{Tr}(ab)}\\
&= (q-1)\sum_{\mathrm{Tr}_{E/F}(a)=0}\chi(a)\xi_p^0 + \sum_{\mathrm{Tr}_{E/F}(a)\neq 0}\chi(a)\sum_{b\in F^*}\xi_p^{\mathrm{Tr}(b\mathrm{Tr}_{E/F}(a))}\\
&= (q-1)\chi(D) - \sum_{\mathrm{Tr}_{E/F}(a)\neq 0}\chi(a)\\
&= (q-1)\chi(D) + \sum_{\mathrm{Tr}_{E/F}(a)=0}\chi(a)\\
&= q\chi(D).
\end{aligned}$$

This proves the lemma. □

The following well known result of Stickelberger [124] completely determines the factorization of Gauss sums. A proof can be found in [130, Prop. 6.13]. For the formulation of Stickelberger's theorem, we need some preparations. Let p be a prime, and let $q = p^a$ be a power of p. For every prime ideal π of $\mathbb{Q}(\xi_{q-1})$ above p, one can define the corresponding **Teichmüller character** $\omega(\pi)$ of $\mathbb{F}_q^*$, see [130, p. 96]. For our purposes, the explicit definition of $\omega(\pi)$ is not relevant, we only need to know that ω has order $q-1$ and thus generates the character group of $\mathbb{F}_q^*$. For a positive integer x, let $S_p(x)$ denote the sum of the p-digits of x.

Theorem 1.4.9 (Stickelberger) *Let p be a prime, and let $q = p^a$ be a power of p. Let π be a prime ideal of $\mathbb{Q}(\xi_{q-1})$ above p, let $\tilde{\pi}$ be the prime ideal of $\mathbb{Q}(\xi_{q-1}, \xi_p)$ above π. By $\nu_{\tilde{\pi}}$ we denote the $\tilde{\pi}$-adic evaluation. Let $\omega = \omega(\pi)$ be the Teichmüller character of $\mathbb{F}_q^*$ corresponding to π. Then*

$$\nu_{\tilde{\pi}}(G(\omega^j)) = S_p(j)$$

for $1 \le j < q-1$.

We also need the Davenport-Hasse Theorem, see [76, Thm. 5.14], which we recall in the following.

Theorem 1.4.10 (Davenport-Hasse) *Let r be a prime power, and let s be a positive integer. Write $F := \mathbb{F}_r$ and $E = \mathbb{F}_{r^s}$. Let χ be a character of F^* and define a character χ' of E^* by $\chi'(x) = \chi(\mathrm{N}_{E/F}(x))$ where $\mathrm{N}_{E/F}$ denotes the norm of E relative to F. Then*

$$G(\chi') = (-1)^{s-1}G(\chi)^s.$$

Finally, we state an elementary lemma on the structure of the group of units of $\mathbb{Z}_{p^a}$. It is implicitly contained in [62, pp. 274-276], for instance.

Lemma 1.4.11 *Let p be a prime, and let b be a positive integer.*
a) Assume $(p,b) \neq (2,1)$. If s is an integer satisfying $s \equiv 1 \pmod{p^b}$ and $s \not\equiv 1 \pmod{p^{b+1}}$ then $\mathrm{ord}_{p^c}(s) = p^{c-b}$ for all $c \geq b$.
b) Let s and t be integers such that $\mathrm{ord}_{p^b}(s) = \mathrm{ord}_{p^b}(t)$ is a power of p. Furthermore, assume $s \equiv t \equiv 1 \pmod 4$ if $p = 2$. Then s and t generate the same subgroup of the multiplicative group $\mathbb{Z}_{p^b}^*$.

Note that the assumption $(p,b) \neq (2,1)$ in Lemma 1.4.11 a) is necessary since, for instance, $\mathrm{ord}_8(3) = 2 \neq 2^{3-1}$. The assumption $s \equiv t \equiv 1 \pmod 4$ in part b) also is essential. For instance, $\mathrm{ord}_8(3) = \mathrm{ord}_8(5) = 2$, but 3 and 5 generate different subgroups of $\mathbb{Z}_8^*$.

1.5 Algebraic-combinatorial tools

In the last section, we provided some tools for the first main step of the algebraic approach to difference sets and similar objects, namely, getting information on the solutions of $X\overline{X} = n$ in cyclotomic integers. In the present section, we supply some tools for retranslating this information into necessary conditions on putative difference sets, the second step of the algebraic approach. We begin with an extremely useful result due to Ma [82], see also [12, VI, Cor. 13.5] or [105, Cor. 1.2.14].

Lemma 1.5.1 (Ma) *Let p be a prime and let G be a finite abelian group with a cyclic Sylow p-subgroup S. If $Y \in \mathbb{Z}[G]$ satisfies*

$$\chi(Y) \equiv 0 \bmod p^a$$

for all characters χ of G of order divisible by $|S|$, then there exist $X_1, X_2 \in \mathbb{Z}[G]$ such that

$$Y = p^a X_1 + P X_2,$$

where P is the unique subgroup of order p of G. Furthermore, if Y has nonnegative coefficients only, then X_1 and X_2 also can be chosen to have nonnegative coefficients only.

Sometimes, it is necessary to use the following refinement of Ma's lemma. See [3] for a proof.

Lemma 1.5.2 *Let p be a prime, and let G be an abelian group with a cyclic Sylow p-subgroup of order p^s. If $Y \in \mathbb{Z}[G]$ satisfies*

$$\chi(Y) \equiv 0 \pmod{p^a}$$

for all characters χ *of* G, *then there are* $X_0, X_1, ..., X_r \in \mathbb{Z}[G]$ *with*

$$Y = p^a X_0 + p^{a-1} P_1 X_1 + \cdots + p^{a-r} P_r X_r$$

where $r = \min\{a, s\}$ *and* P_i, $i = 1, ..., r$, *is the subgroup of order* p^i *of* G *(viewed as an element of* $\mathbb{Z}[G]$*).*

Finally, we state a lemma from [84] which will be needed in Section 3.2.6 for the proof of a unified exponent bound for relative difference sets with prime power parameters.

Lemma 1.5.3 *Let* P *be a cyclic group of order* p^t *where* p *is an odd prime, and let* P_i *be the unique subgroup of order* p^i *of* P $(0 \le i \le t)$. *If* $A \in \mathbb{Z}[P]$ *satisfies*

$$\chi(A)\overline{\chi(A)} = p^{2c}$$

for all $\chi \in P^* \setminus P_n^\perp$, *where* $1 \le n \le t$ *and* $n \le c+1$, *then*

$$A = \sum_{m=0}^{n-1} \epsilon_m (p^{c-m} P_m - p^{c-m-1} P_{m+1}) g_m + P_n Y$$

with $\epsilon_m = \pm 1$, $g_m \in P$ *and* $Y \in \mathbb{Z}[P]$.

Chapter 2

The field descent

2.1 The fixing theorem

All that can be said *a priori* about the character sums $\chi(X)$ corresponding to combinatorial structures such as difference sets, planar functions or group invariant weighing matrices is that $\chi(X)$ is an algebraic integer of a prescribed absolute value in the eth cyclotomic field $\mathbb{Q}(\xi_e)$ over the rationals where e is the order of χ.
In this section, we will show that in most cases one can say much more, namely, that $\chi(D)$ times a root of unity lies in a "small" subfield of $\mathbb{Q}(\xi_e)$. The exact formulation of this this **"field descent"** will be given in Theorems 2.2.8, 2.2.2, 2.2.3. The idea of the field descent can be generalized to arbitrary CM-fields, see Section 2.4.2. One of the basic questions concerning the field descent is the following.

Problem 2.1.1 *Consider a principal ideal I of a number field K and assume that I is invariant under a certain automorphism σ of K. Is there a generator g of I with $g^\sigma = g$?*

We will see in Section 2.4 that Problem 2.1.1 is closely related to the structure of the class group of K. In the case where I is generated by a solution of the modulus equation 1.7, we can use Kronecker's lemma 1.4.1 to find simple sufficient conditions for a positive answer to Problem 2.1.1 for CM-fields. Among other things, these conditions will be given in the crucial Theorem 2.1.4. First we state some preparatory lemmas. We use the notation $\xi_t := \exp 2\pi i/t$ and write $\mathrm{ord}_a(b)$ for the multiplicative order of b modulo a.

Lemma 2.1.2 *Let $X \in \mathbb{Z}[\xi_m]$ be a solution of $X\overline{X} = n$ where n is a positive integer. If $\sigma \in \mathrm{Gal}(\mathbb{Q}(\xi_m)/\mathbb{Q})$ fixes all primes above n in $\mathbb{Q}(\xi_m)$, then*

$$X^\sigma = \varepsilon X$$

for some root of unity ε.

Proof Since $(X) = (X^\sigma)$, we have $X^\sigma = uX$ for some unit u. As $|X^\sigma|^2 = (X\overline{X})^\sigma = n^\sigma = n = |X|^2$, u has absolute value 1. Now Kronecker's lemma 1.4.1 shows that u is a root of unity. □

Lemma 2.1.2 indicates that the following variation of Problem 2.1.1 is important. Later we will see that it is really important.

Problem 2.1.3 (Fixing Problem) *Let $\sigma \in \mathrm{Gal}(\mathbb{Q}(\xi_m)/\mathbb{Q})$ and let*

$$T := \{X \in \mathbb{Q}(\xi_m) : X^\sigma = \varepsilon X \text{ for some root of unity } \varepsilon\}.$$

Is there, for every $X \in T$, a root of unity $\delta_X \in \mathbb{Q}(\xi_m)$ such that $X\delta_X$ remains fixed by σ?

The answer to the Fixing Problem is not always positive. For example, let p be an odd prime and consider a Gauss sum $G(\chi) = \sum_{x\in\mathbb{F}_p^*} \chi(x)\xi_p^x$ where χ is a nontrivial character of $\mathbb{F}_p^*$. Let $\sigma \in \mathrm{Gal}(\mathbb{Q}(\xi_{p(p-1)})/\mathbb{Q})$ be defined by $\sigma(\xi_{p-1}) = \xi_{p-1}$ and $\sigma(\xi_p) = \xi_p^a$ where a is a primitive root modulo p. Then

$$\begin{aligned} G(\chi)^\sigma &= \sum_{x\in\mathbb{F}_p^*} \chi(x)\xi_p^{ax} \\ &= \sum_{x\in\mathbb{F}_p^*} \chi(a^{-1}x)\xi_p^{x} \\ &= \chi(a^{-1})G(\chi). \end{aligned}$$

If the answer to the Fixing Problem was positive for the chosen σ, then $(G(\chi)\xi_{p-1}^i)^\sigma = \chi(a^{-1})G(\chi)\xi_{p-1}^i = G(\chi)\xi_{p-1}^i$ for some i. But this is impossible since $\chi(a^{-1}) \neq 1$.
The following theorem in particular provides sufficient conditions for a positive answer to the fixing problem. In many cases, our conditions are also necessary as can be seen through the example of Gauss sums. One look at these conditions shows that they are messy. However, these are just numerical conditions which are easy to check for any given instance. As we will see in Section 2.4 and Chapter 3, this result is important for the study of class groups of CM-fields as well as for the algebraic approach to difference sets.

For a prime p and an integer x, let x_p be the p-part of x, i.e., $x = x_p x'$ where x_p is a power of p and $(x', p) = 1$.

Theorem 2.1.4 (Fixing Theorem) *Let K/k be a Galois extension of algebraic number fields. Let $\mathbb{Q}(\xi_m)$, $m \not\equiv 2 (\mathrm{mod}\ 4)$, be the largest cyclotomic field contained in K. Let $\sigma \in \mathrm{Gal}(K/k)$, and denote the order of σ by y. Define t by $\sigma(\xi_m) = \xi_m^t$. Write $m = \prod_{p\in S} m_p$ where S is the set of primes dividing m. Let*

$$T_{odd} := \{p \in S : p \text{ odd},\ t \equiv 1 \pmod p),\ y_p > \mathrm{ord}_{m_p}(t)\}$$

and

$$T := \begin{cases} T_{odd} \cup \{2\} & \text{if } t \equiv 1 \pmod 4 \text{ and } y_2 > \text{ord}_{m_2}(t), \\ T_{odd} & \text{otherwise.} \end{cases}$$

Define

$$f(m,\sigma) := \begin{cases} 2\gcd(m, \prod_{p\in T} y_p) & \text{if } m \text{ is odd and } y \text{ is even,} \\ m_2 \gcd(m, \prod_{p\in T} y_p) & \text{if } m \text{ is even, } t \equiv 3 \pmod 4 \\ & \text{and } 2m_2 \text{ divides } t^y - 1, \\ \gcd(m, \prod_{p\in T} y_p) & \text{otherwise.} \end{cases}$$

If

$$X^\sigma = \varepsilon X \tag{2.1}$$

for $X \in K$ and some root of unity ε, then there is an mth root of unity α and an $f(m,\sigma)$th root of unity η with

$$(X\alpha)^\sigma = \eta(X\alpha).$$

In particular, $(X\alpha)^{f(m,\sigma)} \in \text{Fix}\ \sigma$.

Proof Write $\varepsilon = \delta \prod_{q\in S} \lambda_p$ where each λ_p is m_pth root of unity, $\delta = \pm 1$ and $\delta = 1$ if m is even. We apply σ to (2.1) repeatedly $y-1$ times and get $\varepsilon^{(t^y-1)/(t-1)} = 1$.
This implies

$$\delta^y = 1, \tag{2.2}$$

and

$$\lambda_p^{\frac{t^y-1}{t-1}} = 1 \tag{2.3}$$

for all $p \in S$.

Claim 1: If $p \in S$ is odd and $t \equiv 1 \pmod p$ or $p = 2$ and $t \equiv 1 \pmod 4$, then

$$y_p \ || \ \frac{t^y - 1}{t-1}. \tag{2.4}$$

Proof of Claim 1: Define b by $p^b \ || \ t-1$. By Lemma 1.4.11 a, we have $\text{ord}_{y_p p^b}(t^y) = \text{ord}_{y_p p^b}(t)/y_p = y_p/y_p = 1$ and $\text{ord}_{y_p p^{b+1}}(t^y) = p$. Thus $y_p p^b \ || \ t^y - 1$ and the claim follows.

Claim 2: Let $p \in S \setminus T$. If p is odd or $p = 2$ and $t \equiv 1 \pmod 4$, then there is a solution i_p to

$$\xi_{m_p}^{i_p(t-1)} = \lambda_p. \tag{2.5}$$

Proof of Claim 2: If $t \not\equiv 1 \pmod p$, then (2.5) certainly has a solution since λ_p is a m_pth root of unity. Thus we may assume $t \equiv 1 \pmod p$. Then $\mathrm{ord}_{m_p}(t)$ is a power of p and thus divides y_p. Using the definition of T, we conclude $\mathrm{ord}_{m_p}(t) = y_p$. If $y_p = 1$, then by (2.3) and (2.4), we get $\lambda_p = 1$ and (2.5) has a solution. Thus we may assume $y_p > 1$. Again define b by $p^b \mid\mid t-1$. Since $t \equiv 1 \pmod p$ and $\mathrm{ord}_{m_p}(t) = y_p > 1$, we have $p \le p^b < m_p$. From Lemma 1.4.11 we infer $y_p = \mathrm{ord}_{m_p}(t) = m_p/p^b$. Thus, by (2.3) and (2.4), we get $\lambda_p^{m_p/p^b} = 1$. This shows that (2.5) has a solution since $\xi_{m_p}^{(t-1)}$ is a primitive p^{m_p/p^b}th root of unity. Thus Claim 2 is proven.

Claim 3: Let m be even, $t \equiv 3 \pmod 4$ and assume that $2m_2$ does not divide $t^y - 1$. Then (2.5) has a solution for $p = 2$.
Proof of Claim 3: Since m_2 divides $t^y - 1$, we have $m_2 || t^y - 1$. Since $2 \mid\mid t-1$, we get $(m_2/2) \mid\mid (t^y - 1)(t-1)$. Thus, by (2.3) and (2.4), we get $\lambda_2^{m_2/2} = 1$. Thus (2.5) has a solution since $\xi_{m_2}^{(t-1)}$ is a primitive $2^{m_2/2}$th root of unity. This proves Claim 3.

Claim 4: Let U be the set of primes p in S for which (2.5) has a solution i_p. Let $\gamma := \prod_{p\in U} \xi_{m_p}^{i_p}$. Then

$$(X\gamma)^\sigma = \left(\delta \prod_{p\in S\setminus U} \lambda_p\right)(X\gamma) \tag{2.6}$$

Proof of Claim 4: This is a straightforward calculation using $\varepsilon = \delta \prod_{q\in S} \lambda_p$ and (2.5).

Claim 5: $\omega := \delta \prod_{p\in S\setminus U} \lambda_p$ is an $f(m,\sigma)$th root of unity.
Proof of Claim 5: First let p be odd. If $p \in S \setminus U$, then $p \in T$ by Claim 2. Furthermore, by (2.2) and (2.3), we get $\lambda_p^{y_p} = 1$. Thus, by the definition of $f(m,\sigma)$, we have

$$\left(\prod_{p\in S\setminus\{U\cup\{2\}\}} \lambda_p\right)^{f(m,\sigma)} = 1.$$

Now consider $p = 2$. If $2 \in S \setminus U$, then by Claims 3, 4 we have $2 \in T$ and $t \equiv 1 \pmod 4$ or $t \equiv 3 \pmod 4$ and $2m_2$ divides $t^y - 1$. In both cases, the definition of $f(m,\sigma)$ makes sure that $\lambda_2^{f(m,\sigma)} = 1$.
Summing up, we have shown $(\omega\delta)^{f(m,\sigma)} = 1$. It remains to show $\delta^{f(m,\sigma)} = 1$. For even m we have $\delta = 1$, i.e., in this case, there is nothing to show. Let m be odd. If y is odd, too, then $\delta = 1$ by (2.2) and we are done. If y is even, then $f(m,\sigma)$ is even by definition and thus $\delta^{f(m,\sigma)} = 1$. This proves Claim 5.

Conclusion of the proof: The assertion of the Theorem follows from Claims 4 and 5. We take $\alpha = \gamma$ and $\eta = \omega$. □

2.2 Prescribed absolute value

As a first application of Theorem 2.1.4, we are now going to prove a necessary condition on solutions of the modulus equation $X\overline{X} = n$ in the case where $n = p^a$ for a rational prime p. In a sense, this result will tell us that any "ramified" solution of $X\overline{X} = p^a$ necessarily is a Gauss sum times an "unramified" solution.

Notation 2.2.1 Let $p = ef + 1$ be an odd prime where $e \neq 1$. Let Γ be the set of all characters of $\mathbb{Z}_p^*$ of order dividing e. We define

$$G(p,e) := \{G(\chi) : \chi \in \Gamma\}.$$

For the definition of the Gauss sums $G(\chi)$, see Definition 1.4.6. For $p = 2$, we define $G(2,2) := \{1 + i\}$.

Theorem 2.2.2 *Let $m = p^a m'$ where p is a prime, $(p, m') = 1$ and $m \not\equiv 2 \pmod 4$. If $X \in \mathbb{Z}[\xi_m]$ is a solution of $X\overline{X} = p^b$, $b \geq 1$, then there is an integer j such that*

$$X\xi_m^j \in \mathbb{Z}[\xi_{m'}] \text{ or } X = \xi_m^j G(\chi) Z,$$

where $Z \in \mathbb{Z}[\xi_{m'}]$, $Z\overline{Z} = p^{b-1}$ and $G(\chi) \in G(p, w_0)$ with

$$w_0 = \begin{cases} 2 & \text{if } p = 2, \\ (p-1, m') & \text{if } m' \text{ is even,} \\ 2(p-1, m') & \text{if both } p \text{ and } m' \text{ are odd.} \end{cases}$$

Proof The proof mainly relies on Theorem 2.1.4, and we will use the function $f(m, \sigma)$ defined there. Write $m' = \prod_{i=1}^{s} q_i^{a_i}$ where the q_i are distinct primes.
a) We first consider the case $p = 2$. Let t be an integer satisfying $t \equiv 5 \pmod{2^a}$ and $t \equiv 1 \pmod{m'}$ and let $\sigma \in \mathrm{Gal}(\mathbb{Q}(\xi_m)/\mathbb{Q})$ be defined by $\sigma(\xi_m) = \xi_m^t$. Note $y := \mathrm{ord}(\sigma) = \mathrm{ord}_m(t) = 2^{a-2}$. From Theorem 1.4.3 we know that σ fixes all primes above 2 in $\mathbb{Z}[\xi_m]$. Thus $X^\sigma = \varepsilon X$ for some root of unity ε by Lemma 2.1.2. Note that $T = \emptyset$ and $f(m, \sigma) = 1$ in Theorem 2.1.4 for our σ here. Thus $X_1 := X\xi_m^r \in \mathrm{Fix}\,\sigma \cap \mathbb{Z}[\xi_m] = \mathbb{Z}[\xi_{4m'}]$ for some integer r. Let σ_1 be defined by $i \to -i$, $\xi_{m'} \to \xi_{m'}$. By Theorem 1.4.3 and Theorem 2.1.4 there are $r_1, r_2 \in \{0,1,2,3\}$ such that $X_2^{\sigma_1} = i^{r_1} X_2$ where $X_2 := X_1 \xi_{m'}^{r_2}$. Write $X_2 = y_1 + y_2 i$ with $y_1, y_2 \in \mathbb{Z}[\xi_{m'}]$. If $r_1 = 0$ then $X_2 \in \mathrm{Fix}\,\sigma_1$ and we are finished. If $r_1 = 2$ then $X_2^{\sigma_1} = y_1 - y_2 i = -X_2 = -y_1 - y_2 i$. Hence $X_2 = y_2 i$ yielding the assertion. If $r_1 = 1$ then $X_2^{\sigma_1} = y_1 - y_2 i = iX_2 =$

$-y_2 + y_1 i$. Thus $y_1 = -y_2$, i.e., $X_2 = (1-i)y_1 = -i(1+i)y_1$ again yielding the assertion. The case $r_1 = 3$ is similar. This completes the proof for $p = 2$.
b) Let p be odd. We first show $X\xi_m^j \in \mathbb{Z}[\xi_{pm'}]$ for some j. Let t be an integer satisfying $t \equiv p+1 \pmod{p^a}$ and $t \equiv 1 \pmod{m'}$. Then, by Theorem 1.4.3, $\sigma \in \mathrm{Gal}(\mathbb{Q}(\xi_m)/\mathbb{Q})$ defined by $\sigma(\xi_m) = \xi_m^t$ fixes all primes above p. Note that $f(m,\sigma) = 1$ in Theorem 2.1.4 for the chosen σ. Thus we get $X_1 := X\xi_m^j \in \mathrm{Fix}\,\sigma \cap \mathbb{Z}[\xi_m] = \mathbb{Z}[\xi_{pm'}]$ for some j.
Now choose an integer t_1 with $t \equiv 1 \pmod{m'}$ and $t_1 \equiv h \pmod{p}$ where h is a primitive root modulo p. Let $\sigma_1 \in \mathrm{Gal}(\mathbb{Q}(\xi_{pm'})/\mathbb{Q})$ be defined by $\sigma(\xi_{pm'}) = \xi_{pm'}^{t_1}$. From Theorem 1.4.3 we know that σ_1 fixes all primes above p in $\mathbb{Q}(\xi_{pm'})$. Once more, we have $X^\sigma = \varepsilon X$ for some root of unity ε by Lemma 2.1.2. It is straightforward to check that $f(m,\sigma) = w_0$ where f is the function defined in Theorem 2.1.4 and w_0 is defined above.
It follows that $X_1^{\sigma_1} = \eta X_1$ where η is a w_0th root of unity. If $\eta = 1$, then $X_1 \in \mathrm{Fix}\,\sigma_1 \cap \mathbb{Z}[\xi_m] = \mathbb{Z}[\xi_{m'}]$ yielding the assertion. Thus assume $\eta \neq 1$. We write $X_1 = \sum_{i=0}^{p-2} A_i \xi_p^{h^i}$ with $A_i \in \mathbb{Z}[\xi_{m'}]$. Then

$$\begin{aligned} X_1^{\sigma_1} &= \eta \sum_{i=0}^{p-2} A_i \xi_p^{h^i} \\ &= \sum_{i=0}^{p-2} A_i \xi_p^{h^{i+1}} \\ &= A_{p-2}\xi_p^{h^0} + \sum_{i=1}^{p-2} A_{i-1}\xi_p^{h^i}. \end{aligned}$$

Hence $A_0\eta = A_{p-2}$ and $A_i\eta = A_{i-1}$ for $i = 1, ..., p-2$. Thus $A_i = A_0\eta^{-i}$, $i = 1, ..., p-2$. This gives $X_1 = A_0 \sum_{i=0}^{p-2} \eta^{-i}\xi_p^{h^i}$ completing the proof. □

The following theorem gives a restriction on the solutions of $X\overline{X} = n$ of a different type. A special case of this result was proved in [20].

Theorem 2.2.3 *Let $X \in \mathbb{Z}[\xi_m]$ be a solution of $X\overline{X} = n$ where $m = p^a$, p is an odd prime and $(n,p) = 1$ Write $n = \prod_{i=1}^s r_i^{a_i}$ where the r_is are distinct primes. If $a \geq 2$, we assume $r_i^{p-1} \not\equiv 1 \pmod{p^2}$ for all i. Let f be any common divisor of $\mathrm{ord}_p(r_i)$, $i = 1, ..., s$, and write $p = ef + 1$. Then the following hold.*
a) If n is a square of a positive integer u and $f > 2u(p-1)/p$, then $(X) = (u)$.
b) If n is a nonsquare, then f is odd and there is an integer y satisfying $y^2 \equiv e^2 n \pmod{p}$ and $1 \leq y \leq e\sqrt{n}$. In particular, $e^2 n > p$.

Proof If f is even, then p is self-conjugate modulo m. Thus, by Theorem 1.4.3 the primes above p in $\mathbb{Q}(\xi_m)$ are invariant under complex conjugation. Hence the assertion follows from Lemma 1.4.2 and Corollary 1.4.5 in this case.

Now assume that f is odd. Let t be an integer such that $\mathrm{ord}_{p^a}(t) = fp^{a-1}$ and define $\sigma \in \mathrm{Gal}(\mathbb{Q}(\xi_m)/\mathbb{Q})$ by $\xi_m \to \xi_m^t$. The assumptions of Theorem 2.2.3 together with Lemma 1.4.11 imply that fp^{a-1} divides $\mathrm{ord}_{p^a}(r_i)$ for all i. We conclude that for every i there is an integer j_i such that $r_i^{j_i} \equiv t \pmod{p^a}$. Thus, by Theorem 1.4.3, σ fixes all primes above n in $\mathbb{Z}[\xi_m]$.
As $\mathrm{ord}_{p^a}(t) = fp^{a-1}$ is odd and since $(p, t-1) = 1$, we have $f(m,\sigma) = 1$ in Theorem 2.1.4 for the chosen σ. Thus we may assume $X \in K_{p,e}$ where $K_{p,e}$ is the subfield of dimension $e := (p-1)/f$ of $\mathbb{Q}(\xi_p)$. Let g be a primitive root modulo p. The Gaussian periods $\eta_i = \sum_{t=0}^{f-1} \xi_p^{g^{et+i}}$, $i = 0, ..., e-1$, form an integral basis of $K_{p,e}$ over $\mathbb{Q}$. Hence we can write $X = \sum_{i=0}^{e-1} b_i \eta_i$ with $b_i \in \mathbb{Z}$. It is shown in [21], Lemma 2.3, that this implies

$$en = p \sum b_i^2 - f(\sum b_i)^2 \tag{2.7}$$

and $|\sum b_i| \le e\sqrt{n}$. Considering (2.7) modulo p and multiplying by e we conclude $y^2 \equiv e^2 n \pmod{p}$ where $y = |\sum b_i|$. This already proves part b). Furthermore, if $n = u^2$ for a positive integer u, then $y \equiv \pm eu \pmod{p}$. Since $y \le eu$ and $p > 2u(p-1)/f = 2ue$, we infer $y = ue$. Now (2.7) gives $\sum b_i^2 = eu^2$ and this together with $|\sum b_i| = eu$ implies that $b_i = u$ or $b_i = -u$ for all i completing the proof of a). □

Remark 2.2.4 It is possible to prove a version of Theorem 2.2.3 for m being the product of two prime powers. This is quite tedious and straightforward and will not be done here.

Now we come to the most important version of the field descent for cyclotomic integers of prescribed absolute value. It will turn out that the integer $F(m,n)$ defined below describes a subring $\mathbb{Z}[\xi_{F(m,n)}]$ of $\mathbb{Z}[\xi_m]$ that already contains all solutions $X \in \mathbb{Z}[\xi_m]$ of $X\overline{X} = n$ up to multiplication with a root of unity. The prime 2 will need special attention in our considerations as the multiplicative group modulo 2^a is noncyclic for $a \ge 3$.

Definition 2.2.5 *Let m, n be positive integers, and let $m = \prod_{i=1}^t {p_i}^{c_i}$ be the prime power decomposition of m. For each prime divisor q of n let*

$$m_q := \begin{cases} \prod_{p_i \ne q} p_i & \text{if } m \text{ is odd or } q = 2, \\ 4\prod_{p_i \ne 2,q} p_i & \text{otherwise.} \end{cases}$$

Let $\mathcal{D}(n)$ be the set of prime divisors of n. We define $F(m,n) = \prod_{i=1}^t {p_i}^{b_i}$ to be the minimum multiple of $\prod_{i=1}^t p_i$ such that for every pair (i,q), $i \in \{1,...,t\}$, $q \in \mathcal{D}(n)$, at least one of the following conditions is satisfied.
(a) $q = p_i$ and $(p_i, b_i) \ne (2,1)$,
(b) $b_i = c_i$,
(c) $q \ne p_i$ and $q^{\mathrm{ord}_{m_q}(q)} \not\equiv 1 \pmod{p_i^{b_i+1}}$.

Remark 2.2.6 For the sake of clarity, we also provide an explicit formula for the numbers b_i. First note that, for fixed i, the set of positive integers x satisfying

$$q^{\mathrm{ord}_{m_q}(q)} \not\equiv 1 \pmod{p_i^{x+1}}$$

for all $q \in \mathcal{D}(n) \setminus \{p_i\}$ is a ray $[e_i, \infty)$ with $e_i \geq 1$. We have

$$b_i = \begin{cases} 2 & \text{if } p_i = 2,\ c_i \geq 2 \text{ and } e_i = 1, \\ \min(c_i, e_i) & \text{otherwise.} \end{cases}$$

The reason why $b_i = 2$ if $p_i = 2$, $e_i = 1$ and $c_i \geq 2$ is the following. Note that n must be a power of 2 if $p_i = 2$ and $e_i = 1$ (if n has an odd prime divisor q, then $q^{\mathrm{ord}_{m_q}(q)} \equiv 1 \pmod 4$ by the definition of m_q since m is even if $p_i = 2$). If $p_i = 2$ and n is a power of 2, then (a) or (b) must hold for $p_i = q = 2$, and the condition $(p_i, b_i) \neq (2,1)$ in (a) makes sure that $b_i = 2$ if $c_i \geq 2$. Also note that we have $b_i \geq 2$ if $p_i = 2$ and $m \not\equiv 2 \pmod 4$.

It is worth to note the following important property of $F(m,n)$.

Proposition 2.2.7 *Let P be a finite set of primes, and let Q be the set of all positive integers which are products of powers of primes in P. Then there is a computable constant $C(P)$ such that*

$$F(m,n) \leq C(P)$$

for all $m, n \in Q$.

Now we are ready to prove the main result of this section.

Theorem 2.2.8 *Assume $X\overline{X} = n$ for $X \in \mathbb{Z}[\xi_m]$ where n and m are positive integers. Then*

$$X\xi_m^j \in \mathbb{Z}[\xi_{F(m,n)}]$$

for some j.

Proof Throughout this proof, we use the notation introduced in Definition 2.2.5. Since $\mathbb{Z}[\xi_m] = \mathbb{Z}[\xi_{m/2}]$ for $m \equiv 2 \pmod 4$, we may assume $m \not\equiv 2 \pmod 4$. Write $m = \prod_{i=1}^t p_i{}^{c_i}$ and $F(m,n) = \prod_{i=1}^t p_i{}^{b_i}$ as in Definition 2.2.5. Recall that $b_i \geq 1$ by definition. Furthermore, $b_i \geq 2$ if $p_i = 2$, see Remark 2.2.6. Let s be an integer satisfying

$$s \equiv 1 \pmod{p_i^{b_i}} \text{ and } s \not\equiv 1 \pmod{p_i^{b_i+1}}$$

for $i = 1, ..., t$. Then $\mathrm{ord}_{p_i^{c_i}}(s) = p_i^{c_i - b_i}$ by Lemma 1.4.11 a. We define $\sigma \in \mathrm{Gal}(\mathbb{Q}(\xi_m)/\mathbb{Q})$ by $\sigma_i(\xi_m) = \xi_m^s$. Note that Fix $\sigma = \mathbb{Q}(\xi_{F(m,n)})$.

Claim: For every prime divisor q of n, the automorphism σ fixes all prime ideals above q in $\mathbb{Z}[\xi_m]$.

Proof of the claim: Write $M_q = \prod_{p_j \neq q} p_j^{c_j}$. By Theorem 1.4.3, our claim is proven if we can show that s is a power of q modulo M_q. Write $Q := q^{\mathrm{ord}_{m_q}(q)}$. Note that $\mathrm{ord}_{p_j^{c_j}}(Q)$ is a power of p_j for all $p_j \neq q$ since $Q \equiv 1 \pmod{p_j}$ by the definition of m_q. We now show that $\mathrm{ord}_{p_i^{c_j}}(Q)$ is divisible by $p_j^{c_j - b_j}$ for all j with $p_j \neq q$. This is trivial if $c_j = b_j$. Otherwise condition (c) of Definition 2.2.5 is satisfied and thus $Q \not\equiv 1 \pmod{p_j^{b_j+1}}$. Then Lemma 1.4.11 a shows that $\mathrm{ord}_{p_j^{c_j}}(Q)$ indeed is divisible by $p_j^{c_j - b_j}$. Note that we need $b_j \geq 1$ for all j and $b_j \geq 2$ for $p_j = 2$ here in order to apply Lemma 1.4.11 a.
Note that the order of Q modulo $p_j^{c_j}$ is a power of p_j for all $p_j \neq q$ and since $Q \equiv 1 (\mathrm{mod}\ 4)$ if M_q is even by Definition 2.2.5. Hence, since $\mathrm{ord}_{p_j^{c_j}}(Q) \geq p_j^{c_j - b_j} = \mathrm{ord}_{p_j^{c_j}}(s)$, Lemma 1.4.11 b implies that s is a power of Q and thus also of q modulo M_q. This proves the claim.

Conclusion of the proof: By our claim and Lemma 2.1.2, we have $X^\sigma = \varepsilon X$ for some root of unity ε. It is straightforward to check that $f(m, \sigma) = 1$ where f is the function defined in Theorem 2.1.4. The assertion now follows from Theorem 2.1.4 since Fix $\sigma = \mathbb{Q}[\xi_{F(m,n)}]$. □

Remark 2.2.9 Note that "the best we can hope for" in Theorem 2.2.8 is $F(m, n) = m_0$ where $m_0 := \prod_{i=1}^{t} p_i$. The worst that can happen is $F(m, n) = m$. As the integers b_i from the definition of $F(m, n)$ have to satisfy quite a lot of conditions of the form

$$q^{\mathrm{ord}_{m_q}(q)} \not\equiv 1 \pmod{p_i^{b_i+1}}$$

it may seem that $F(m, n)$ is usually much bigger than m_0. In order to understand the significance of Theorem 2.2.8 it is important to note that the opposite is the case. Therefore, we consider the following heuristic argument. We assume $n \leq m$ and $m_0 \approx \sqrt{m}$ which is the case for many applications we have in mind ("$\approx$" is not used in a strict sense here). Our claim is that $F(m, n) \approx m_0$ in almost all cases.
To see this we estimate the "probability" that one of the conditions

$$q^{\mathrm{ord}_{m_q}(q)} \not\equiv 1 \pmod{p_i^2} \qquad (*)$$

is violated for a "large" prime divisor p_i of m. Note that $q^{\mathrm{ord}_{m_q}(q)} \equiv 1 \pmod{p_i}$ by the definition of m_q. Furthermore, the probability that p_i divides $\mathrm{ord}_{m_q}(q)$ is very low if p_i is large. So $q^{\mathrm{ord}_{m_q}(q)}$ should take each of the p_i values $1,\ p_i+1, \ldots,\ (p_i-1)p_i+1$ modulo p_i^2 roughly with the same probability. In particular, the probability that $(*)$ is violated should be around $1/p_i$.
Except for a set of density zero all positive integers x have approximately $\log\log x$ prime divisors [54, Thm. 436]. Note $\log\log n \leq \log\log m \approx \log\log m_0^2$

$= \log 2 + \log\log m_0 \approx \log\log m_0$. Thus we usually do not have more than around $(\log\log n)(\log\log m_0) \approx (\log\log m_0)^2$ of the conditions $(*)$.
Now fix any large p_i, say $p_i \approx m_0^{1/\log\log m_0}$ or larger. By the argument above, the probability that $(*)$ is violated for p_i and some fixed q should be around $1/p_i \le 1/m_0^{1/\log\log m_0}$. Hence the probability that *any* of the $\approx (\log\log m_0)^2$ conditions $(*)$ is violated for any large p_i should be less than $(\log\log m_0)^2/m_0^{1/\log\log m_0} = y^2 \exp(-\frac{1}{y}\exp y)$ where $y := \log\log m_0$. Thus, for large m_0, the condition $(*)$ should hold for *all* large p_i and all q with very high probability. By Definition 2.2.5 this amounts to $F(m,n) \approx m_0$.

2.3 Bounding the absolute value

In this section, we will use Theorem 2.2.8 to obtain an upper bound on the absolute value of cyclotomic integers which will be basic for most results of Chapter 3. As a preparation, we need a simple lemma on conjugate characters.
Two characters χ and τ of order e of an abelian group G are called **conjugate** if there is $\sigma \in \mathrm{Gal}(\mathbb{Q}(\xi_e)/\mathbb{Q})$ with $\chi(g) = \tau(g)^\sigma$ for all $g \in G$. Let φ denote the Euler totient function. Though following is well known, see [96, p.6] for instance, we give the quick proof.

Lemma 2.3.1 *Let χ be a character of order e of an abelian group G. Then χ has exactly $\varphi(e)$ distinct conjugates. Futhermore, if $\chi(A) \in \mathbb{Q}$ for some $A \in \mathbb{Z}[G]$, then $\tau(A) = \chi(A)$ for all conjugates τ of χ.*

Proof Since $|\mathrm{Gal}(\mathbb{Q}(\xi_e)/\mathbb{Q})| = \varphi(e)$, we know that χ can have at most $\varphi(e)$ conjugates. Let $\sigma \in \mathrm{Gal}(\mathbb{Q}(\xi_e)/\mathbb{Q})$, $\sigma \neq \mathrm{id}$. Then $\xi_e^\sigma \neq \xi_e$. Moreover, since χ has order e, we can choose $g \in G$ with $\chi(g) = \xi_e$. Then $\chi^\sigma(g) = \xi_e^\sigma \neq \xi_e = \chi(g)$ and thus $\chi^\sigma \neq \chi$. This shows that χ has exactly $\varphi(e)$ conjugates. The second assertion of Lemma 2.3.1 follows from the fact that each $\sigma \in \mathrm{Gal}(\mathbb{Q}(\xi_e)/\mathbb{Q})$ fixes $\mathbb{Q}$ elementwise. □

The following theorem on cyclotomic integers of prescribed absolute value is the most important result of this monograph.

Theorem 2.3.2 (F-bound) *Let $X \in \mathbb{Z}[\xi_m]$ be of the form*

$$X = \sum_{i=0}^{m-1} a_i \xi_m^i$$

with $0 \le a_i \le C$ for some constant C and assume that $n := X\overline{X}$ is an integer. Then

$$n \le \frac{C^2 F(m,n)^2}{4\varphi(F(m,n))}.$$

Proof By Theorem 2.2.8, we can assume $X \in \mathbb{Z}[\xi_f]$ where $f := F(m,n)$. Note that by definition, f is divisible by every prime divisor of m. Thus $1, \xi_m, ..., \xi_m^{m/f-1}$ are independent over $\mathbb{Q}(\xi_f)$. This implies $X = \sum_{i=0}^{f-1} b_i \xi_f^i$ where $b_i := a_{im/f}$. Now we view X also as an element of the group ring $\mathbb{Z}[G]$ where $G = \langle \xi_f \rangle$. Note that $\overline{X} \in \mathbb{Z}[\xi_f]$ correspondes to $X^{(-1)} \in \mathbb{Z}[G]$. Since $X\overline{X} = n \in \mathbb{Q}$, we have

$$\chi(X)\overline{\chi(X)} = n \tag{2.8}$$

for all $\varphi(f)$ characters χ of G of order f by Lemma 2.3.1. Write $l := \sum_{i=0}^{f-1} b_i$. The coefficient of 1 in $XX^{(-1)}$ is $\sum_{i=0}^{f-1} b_i^2$. From the Fourier inversion formula, we get $f \sum_{i=0}^{f-1} b_i^2 = \sum_{\tau \in G^*} |\tau(X)|^2$. Using (2.8) and $\chi_0(X) = l$ for the trivial character χ_0 of G, we get

$$f \sum b_i^2 \geq l^2 + \varphi(f)n. \tag{2.9}$$

Since $0 \leq b_i \leq C$, we have $\sum b_i^2 \leq Cl$. Thus $f \sum b_i^2 - l^2 \leq fCl - l^2 \leq f^2C^2/4$. Combining this with (2.9) gives the assertion. □

2.4 The modulus equation and class groups

In this section, we discuss some applications of the "field descent" for solutions of the modulus equation to number theory. Though these results are only a by-product of the ideas developed in Section 2, they give an interesting and simple approach to some aspects of the structure of class groups of number fields and yield some information that seems hard to get by the usual methods of class field theory. We will use the idea of the field descent to study class groups C of CM-fields. A CM-field is a totally imaginary quadratic extension of a totally real number field, see [130, p. 38]. All cyclotomic fields are CM-fields. We will be able to identify some "interesting" homomorphic images $\overline{C}$ of C. Interesting — since we will be able to obtain lower bounds on the order and the p-ranks of $\overline{C}$. As a consequence, we also get *explicit* generators for subgroups of C and lower bounds on their orders and p-ranks. Our bounds are comparable to bounds that have been obtained by using heavier machinery of class field theory, see [19, 24, 25, 26, 55, 68, 91, 99, 100, 119]. Besides its simplicity, the advantage of our approach is that we are able to obtain *explicit* generators for subgroups of C and in some cases even determine the structure of certain homomorphic images $\overline{C}$ almost completely. The disadvantage of our argument is that it is restricted to CM-fields while the class field theory approach works for arbitrary number fields.

First we will focus on cyclotomic fields since the field descent gives the nicest results for these instances of CM-fields. This is because on the one hand, the factorization of rational primes in cyclotomic fields is completely known, and, on the other hand, we can make use of Theorem 2.2.2 together with Stickelberger's Theorem 1.4.9.

Our main result on class groups of cyclotomic fields will be as follows. Let C be the class group of $\mathbb{Q}(\xi_m)$, and let C^+ be the class group of the maximal real subfield of $\mathbb{Q}(\xi_m)$. Let p be a prime divisor of m, and let p^a be the exact power of p dividing m. By Lemma 1.4.2, we have $(p) = \prod Q_i$ in $\mathbb{Q}(\xi_m)$ where $Q_i = P_i^{\varphi(p^a)}$, and the P_i are prime ideals. Let C_P, C_Q be the subgroups of C generated by the classes $[P_i]$ respectively $[Q_i]$. We will determine the structure of the group C^+C_P/C^+C_Q up to a binary parameter. Another result will provide further explicit bounds on the size of subgroups of C/C^+. Of course, the knowledge of C^+C_P/C^+C_Q yields information on class number factors, a problem which has been studied intensively in the literature — mainly by using class number formulas or class field theory. Work in this direction related to our results can be found in [24, 25, 26, 47, 54, 68, 83, 99, 100].

Finally, we will apply the idea of the field descent to arbitrary CM-fields. This will yield some bounds on orders and p-ranks of their class groups. Our bounds on p-ranks are comparable to, but not as general as those obtained in [119] using heavy number theoretic machinery. Before we turn our attention to cyclotomic fields, we explain the basic idea behind the results of the present section. First we recall some facts on CM-fields, see [130, p. 38].

Lemma 2.4.1 *Let K be a CM-field, i.e., $K = K^+(\sqrt{\alpha})$ where K^+ is totally real and $\alpha \in K^+$ is totally negative.*

a) Complex conjugation induces an automorphism of K which is independent of the imbedding of K in $\mathbb{C}$, i.e.,

$$\alpha^{-1}(\overline{\alpha(x)}) = \beta^{-1}(\overline{\beta(x)})$$

for all $x \in K$ and all imbeddings α, β of K in $\mathbb{C}$.

b) We have $\alpha(\overline{x}) = \overline{\alpha(x)}$ for all $x \in K$ and all imbeddings α of K in $\mathbb{C}$.

c) If ε is a unit in K of modulus 1, then ε is a root of unity.

The following observation clearly shows that the modulus equation over CM-fields is closely related to their class groups.

Proposition 2.4.2 *Let K be a CM-field, let $\mathcal{O}$ be its ring of algebraic integers, and let n be a positive integer. Let I be an ideal of $\mathcal{O}$ solving the ideal equation $I\overline{I} = (n)$. Then the following hold.*

a) The ideal I is principal and has a generator X such that $\varepsilon := n/X\overline{X}$ is a square of a real unit if and only if there is a solution $Y \in \mathcal{O}$ to $Y\overline{Y} = n$ with $(Y) = I$.

b) If I is principal, then there **always** *is a solution $Z \in \mathcal{O}$ of $Z\overline{Z} = n^2$ with $(Z) = I^2$.*

Proof a) If $n/X\overline{X} = \varepsilon^2$ then $Y := \varepsilon X$ solves $Y\overline{Y} = n$ and $(Y) = I$. Conversely, if $Y\overline{Y} = n$ and $(Y) = I$, then $n/Y\overline{Y} = 1$ which surely is a square of a real unit.
b) Let $I = (X)$. The assertion follows from a) since $n^2/X^2\overline{X}^2$ is a square of a real unit. □

Our strategy will be the following. Results like Theorems 2.2.2 and 2.2.3 provide necessary conditions on the ideals (X) generated by solutions of $X\overline{X} = n$. Combined with Proposition 2.4.2 this shows that usually a lot of solutions of the ideal equation $I\overline{I} = (n)$ must be *nonprincipal.* Thus we get a grip on the subgroup of the classgroup generated by the classes of the prime ideals above n.

2.4.1 Class groups of cyclotomic fields

First we are going to utilize Theorem 2.2.2 for the study of class groups of cyclotomic fields. We need some notation.

Notation 2.4.3
Let $m \not\equiv 2 \pmod 4$ be a positive integer, and let $K = \mathbb{Q}(\xi_m)$.

a) Let K^+ be the maximal real subfield of K. By I, H, C, respectively I^+, H^+, C^+, we denote the group of all (fractional) ideals, the group of all principal ideals and the class group of K, respectively K^+. We view I^+, H^+, C^+ as imbedded in I, H, C in the natural way. Note that the imbedding of C^+ in C makes sense, since the natural homomorphism $C^+ \to C$ is an injection, see [130, Theorem 4.14].

b) Write $m = p^a m'$ where p is a prime and $(p, m') = 1$. Recall that p factors in K as $\prod Q_i$, where $Q_i = P_i^{(p-1)p^{a-1}}$ and the P_i are distinct prime ideals. We write I_P, I_Q for the subgroups of I generated by the P_i, respectively Q_i. The groups H_P, H_Q, C_P, C_Q are defined similarly. Thus, for instance, $C_P = I_P H/H$ and $C_Q = I_Q H/H$.

c) Using the notation of Theorem 2.2.2, we define the "Gauss sum group" as the subgroup $G_{m,p}$ of H generated by the ideals (G), $G \in G(p, w_0)$ where

$$w_0 = \begin{cases} 2 & \text{if } p = 2, \\ (p-1, m') & \text{if } m' \text{ is even}, \\ (p-1, 2m') & \text{if both } p \text{ and } m' \text{ are odd}. \end{cases}$$

and $G(p, w_0)$ is the set of all Gauss sums $G(\chi)$ over $\mathbb{F}_p$ for which χ has order dividing w_0.

Lemma 2.4.4 *a) The ideal group*

$$I_P^- := \{J \in I_P : J/\overline{J} \in G_{m,p} I_Q\}$$

contains $I_P \cap I^+ I_Q H$.
b) Assume that p *is not self-conjugate modulo* $m = p^a m'$. *Then*

$$I_P / I_P^- \cong (\mathbb{Z}/w\mathbb{Z}) \times (\mathbb{Z}/u\mathbb{Z})^{e/2-1}$$

where $u = \varphi(p^a)$, $e = \varphi(m')/\mathrm{ord}_{m'}(p)$, $w = u/w_0$ *and* w_0 *is defined above.*

Proof a) Since $I_Q \le I_P^-$ by definition, it suffices to show $I_P \cap I^+ H \le I_P^-$. Thus let $J \in I_P \cap I^+ H$ be arbitrary and write $J = J^+(h)$ with $J^+ \in I^+$ and $h \in K$. Since J^+ is invariant under complex conjugation, we have $J/\overline{J} = (h/\overline{h})$. As $J/\overline{J} \in I_P$, there is a positive integer b such that $y := p^b h/\overline{h} \in \mathbb{Z}[\xi_m]$. Because of $y\overline{y} = p^{2b}$ we conclude $(y) \in G_{m,p} I_Q$ by Theorem 2.5. Hence $J/\overline{J} = (y)/(p^b) \in G_{m,p} I_Q$, too, since $(p^b) \in I_Q$. This shows $I_P \cap I^+ I_Q H \le I_P^-$.

b) Let P_1, $\overline{P_1}$,...,$P_{e/2}$, $\overline{P_{e/2}}$ denote the primes above p in $\mathbb{Z}[\xi_m]$ and define

$$T := \{\prod_{i=1}^{e/2} P_i^{c_i} : 0 \le c_1 \le w-1, 0 \le c_i \le u-1 \text{ for } i = 2, ..., e/2\}.$$

We first show that the elements of T represent distinct cosets of I_P^- in I_P. Thus assume $S := \prod_{i=1}^{e/2} P_i^{c_i - c_i'} \in I_P^-$ with $0 \le c_1, c_1' \le w-1$ and $0 \le c_i, c_i' \le u-1$ for $i = 2, ..., e/2$. Then $S/\overline{S} \in G_{m,p} I_Q$ by the definition of I_P^-. Theorem 2.5 implies that every element of $G_{m,p} I_Q$ can be written in the form $G^\delta J$ with $\delta \in \{0,1\}$, $G \in G(p, w_0)$ and $J \in I_Q$. Thus we can write $S/\overline{S}$ in this form, say $S/\overline{S} = G_S^{\delta_S} J_S$. If $\delta_S = 1$ then by Stickelberger's theorem (see [61, p.209, Theorem 2] or [130, p.98, 1.19]), P_1 occurs in $S/\overline{S} = G_S^{\delta_S} J_S$ to a power $xu/r + yu$, where x, y are integers with $(x, r) = 1$. This implies $|c_1 - c_1'| = |xu/r + yu| \ge u/r \ge w$ which is impossible. Hence $\delta_S = 0$, i.e. $S/\overline{S} \in I_Q$. We conclude $c_i = c_i'$ for all i showing that the elements of T indeed represent distinct cosets of I_P^-.
Our next goal is to show $I_P^- T = I_P$. For that let $J \in I_P$ be arbitrary, say $J = \prod_{i=1}^{e/2} P_i^{a_i} \overline{P_i}^{b_i}$. Because of $P_i \overline{P_i} \in I_P^-$ we may assume $b_i = 0$ for all i. Write $a_1 = z_1 w + z_2$, where z_1, z_2 are integers with $0 \le z_2 < w$. By Stickelberger's theorem there is $G \in G_{m,p}$ such that $G = P_1^w \overline{P_1}^{u-w} \prod_{i=2}^{e/2} P_i^{f_i} \overline{P_i}^{u-f_i}$ for some integers f_i. Then $J_1 := P_1^w \prod_{i=2}^{e/2} P_i^{f_i} \in I_P^-$, since $I_Q J_1 \overline{J_1}^{-1} = I_Q G$. Note that $P_1^{z_2} \prod_{i=2}^{e/2} P_i^{a_i - z_1 f_i}$ can be written as $L_1 L_2$ with $L_1 \in T$ and $L_2 \in I_Q$. Thus $J = P_1^{z_2} J_1^{z_1} \prod_{i=2}^{e/2} P_i^{a_i - z_1 f_i} = (J_1^{z_1} L_2) L_1 \in I_P^- T$. This shows $I_P^- T = I_P$. We conclude $[I_P : I_P^-] = w u^{e/2-1}$. Let U be the subgroup of I_P generated by P_2, $\overline{P_2}$,...,$P_{e/2}$, $\overline{P_{e/2}}$. Then $I_P^- U / I_P^- \cong (\mathbb{Z}/u\mathbb{Z})^{e/2-1}$, since by what we

have shown $I_P^- P_2, ..., I_P^- P_{e/2}$ is a basis of $I_P^- U/I_P^-$. Now assertion b) follows from the theorem on subgroups of free abelian groups of finite rank since the exponent of I_P/I_P^- divides u. □

The following result almost completely determines certain homomorphic images of class groups of cyclotomic fields corresponding to ramified primes.

Theorem 2.4.5 *Assume that p is not self-conjugate modulo $m = p^a m'$. Then*

$$C_P C^+/C_Q C^+ \cong (\mathbb{Z}/2^\delta w\mathbb{Z}) \times (\mathbb{Z}/u\mathbb{Z})^{e/2-1},$$

where $\delta \in \{0,1\}$, $u = \varphi(p^a)$, $e = \varphi(m')/\mathrm{ord}_{m'}(p)$, $w = u/w_0$ where

$$w_0 = \begin{cases} 2 & \text{if } p = 2, \\ (p-1, m') & \text{if } m' \text{ is even}, \\ (p-1, 2m') & \text{if both } p \text{ and } m' \text{ are odd.} \end{cases}$$

In particular, the relative class number h_m^- of $\mathbb{Q}(\xi_m)$ is divisible by $wu^{e/2-1}$.

Proof We first note

$$\begin{aligned} C^+C_P/C^+C_Q &\cong (I^+I_PH/H)/(I^+I_QH/H) \\ &\cong I^+I_PH/I^+I_QH \\ &\cong I_P/I_P \cap I^+I_QH. \end{aligned}$$

Write $A = I_P/I_P \cap I^+I_QH$. The exponent of A divides u and its rank is at most $e/2$. We know from Lemma 3.2 a) that I_P/I_P^- is isomorphic to a factor group of A. Putting these facts together and using Lemma 3.2 b) we see that $A \cong (\mathbb{Z}/v\mathbb{Z}) \times (\mathbb{Z}/u\mathbb{Z})^{e/2-1}$ for some divisor v of u with $v \equiv 0 \pmod{w}$ by the theorem on subgroups of free abelian groups of finite rank.
Our next claim is that $B := I_P^-/I_P \cap I^+I_QH$ is either trivial or an elementary abelian 2-group. For that let $J \in I_P^-$ be arbitrary. Then $J/\overline{J} \in I_P \cap I_QH$ and hence $J^2 = (J/\overline{J})(J\overline{J}) \in I_P \cap I^+I_QH$, as $J\overline{J} \in I_P \cap I^+$. This proves the claim. Finally, since $A/B \cong I_P/I_P^- \cong (\mathbb{Z}/w\mathbb{Z}) \times (\mathbb{Z}/u\mathbb{Z})^{e/2-1}$, we must have $v = w$ or $v = 2w$. □

Remark 2.4.6 *a) If p is self-conjugate modulo $m = p^a m'$ in Theorem 2.4.5 then $C_P C^+/C_Q C^+ = \{1\}$ trivially.*
b) The reason for the binary uncertainty δ in the structure of $C_P C^+/C_Q C^+$ is the loss of information by squaring in Proposition 2.4.2.

It is interesting to compare Theorem 2.4.5 with previously known results which were obtained by different methods. We first consider some results of Metsänkylä [99, 100] who proved congruences for relative class numbers by manipulations of the class number formula. In parts a and b of the following corollary we essentially recover Satz 10 of [99] and in part c) we obtain new congruences which are somewhat related to Satz 8 and Satz 9 of [99].

Corollary 2.4.7 *Let p and q be odd primes and let h_m^- denote the relative class number of $\mathbb{Q}(\xi_m)$. Then the following hold.*
a) $h_{3p^a}^- \equiv 0 \pmod{\varphi(p^a)/6}$ *for* $p \equiv 1 \pmod 3$,
b) $h_{4p^a}^- \equiv 0 \pmod{\varphi(p^a)/4}$ *for* $p \equiv 1 \pmod 4$,
c)

$$h_{p^a q^b}^- \equiv 0 \pmod{\frac{\varphi(p^a)^{(q-1)q^{c-1}/2}}{2q^c}}$$

if q^c, $1 \le c \le b$, *is the highest power of* q *dividing* $p-1$.

Proof a) We put $m' = 3$ in Theorem 2.4.5. Then the assumptions are satisfied and we have $u = \varphi(p^a)$, $e = 2$, $w_0 = 6$, and $w = \varphi(p^a)/6$ implying the assertion.
b) We put $m' = 4$ in Theorem 2.4.5 and get the assertion.
c) Put $m' = q^b$ in Theorem 2.4.5. Then $u = \varphi(p^a)$, $w_0 = 2q^c$ and $e = (q-1)q^{c-1}$, since $\mathrm{ord}_{q^b}(p) = q^{b-c}$. □

Example 2.4.8 We choose an example which can be compared with the table of relative class numbers in [130]. By Corollary 2.4.7 c we have $h_{23\cdot 11}^- \equiv 0 \pmod{2^4 \cdot 11^4}$. The table shows that 2^4, 11^4 are actually the highest powers of 2 respectively 11 dividing $h_{23\cdot 11}^-$.

Another approach to class number factors can be found in [24, 26]; the method is to use Abhyankar's lemma to construct unramified abelian extensions which yield class number factors by class field theory. For instance, it is shown in [24] that the class group of $\mathbb{Q}(\xi_{4p})$ contains a cyclic group of order $(p-1)/4$ if $p \equiv 1 \pmod 4$ is a prime.
Note that this result is contained in our Theorem 2.2.2. A further result from [26] is that the class number of $\mathbb{Q}(\xi_{pq})$ is divisible by $(p-1)/2$ or $(q-1)/2$ if p and q are distinct primes $\equiv 3 \pmod 4$. This is a consequence of the following.

Corollary 2.4.9 *Let p and q be primes $\equiv 3 \pmod 4$. By quadratic reciprocity we may assume that p is a square modulo q, i.e. that $\mathrm{ord}_q(p)$ is odd. Then*

$$h_{pq}^- \equiv 0 \pmod{(p-1)^{(q-1)/(2\mathrm{ord}_q(p))}/2}$$

if $(p-1, q) = 1$ *and*

$$h_{pq}^- \equiv 0 \pmod{(p-1)^{(q-1)/2}/(2q)}$$

if q *divides* $p-1$.

Proof Put $m' = q$ in Theorem 3.3. □

We need some notation for the formulation of our next result. Let m and t be positive, relatively prime integers where $m = p^a$ for an odd prime p.

Furthermore, let $t = \prod_{i=1}^{s} r_i^{d_i}$ be the prime power decomposition of t. We are only interested in the case where $f_i := \mathrm{ord}_p(r_i)$ is odd for every i. Then the prime ideals above r_i in $\mathbb{Z}[\xi_m]$ are <u>not</u> invariant under complex conjugation. Hence each r_i factors in $\mathbb{Z}[\xi_m]$ as

$$(r_i) = \prod_{j=1}^{u_i} P_{ij}\overline{P_{ij}},$$

where $u_i = \varphi(p^a)/(2\mathrm{ord}_{p^a}(r_i))$ and the P_{ij} are distinct prime ideals. We also use Notation 2.4.3.

Theorem 2.4.10 *Assume that f is a common divisor of $f_1,\ldots,f_s$ and that $r_i^{p-1} \not\equiv 1 \pmod{p^2}$ if $m = p^a > p$. If the f_i are odd and if*

$$t = \prod_{i=1}^{s} r_i^{d_i} < \frac{fp}{2(p-1)},$$

where the d_i are any nonnegative integers, then the ideal classes

$$\left[\prod_{i=1}^{s}\prod_{j=1}^{u_i} P_{ij}^{c_{ij}}\right], \quad 0 \le c_{ij} \le d_i,$$

$i = 1,\ldots,s$, $j = 1,\ldots,u_i$, represent distinct cosets of C^+ in C. Here we have $u_i = (p-1)/2f_i$. In particular, the order of the subgroup of C/C^+ generated by the cosets $C^+[P_{ij}]$, $i = 1,\ldots,s$, $j = 1,\ldots,u_i$, is at least $\prod_{i=1}^{s}(d_i+1)^{u_i}$.

Proof Assume $C^+[\prod_{i=1}^{s}\prod_{j=1}^{u_i} P_{ij}^{c_{ij}}] = C^+[\prod_{i=1}^{s}\prod_{j=1}^{u_i} P_{ij}^{c'_{ij}}]$. Then there are $h \in K$ and $J^+ \in I^+$ such that $J := \prod_{i=1}^{s}\prod_{j=1}^{u_i} P_{ij}^{c_{ij}-c'_{ij}} = (h)J^+$. Since $(h/\overline{h}) = J/\overline{J}$ and $|c_{ij} - c'_{ij}| \le d_i$ for all i,j, we know that $y := th/\overline{h}$ lies in $\mathbb{Z}[\xi_m]$. As $y\overline{y} = t^2$ and $f > 2t(p-1)/p$ by the assumption, we can apply Theorem 2.2.3 a and get $(y) = (t)$. Thus $(h) = (\overline{h})$ and $J = \overline{J}$, i.e. $c_{ij} = c'_{ij}$ for all i, j. □

Corollary 2.4.11 *Let $m = p^a$ for an odd prime p. Assume that q is a prime such that $\mathrm{ord}_p(q)$ is odd and that $q^{p-1} \not\equiv 1 \pmod{p^2}$ if $a \ge 2$.*
Then the size of the subgroup of C/C^+ generated by the classes of the primes Q_i above q in $\mathbb{Z}[\xi_m]$ is at least

$$\left(\left\lfloor \ln\frac{\mathrm{ord}_p(q)p}{2(p-1)}/\ln q\right\rfloor + 1\right)^{\frac{p-1}{2\mathrm{ord}_p(q)}}.$$

Furthermore, each Q_i has order at least $\left\lfloor \ln\frac{\mathrm{ord}_p(q)p}{2(p-1)}/\ln q\right\rfloor + 1$ in C.

Proof We put $s = 1$, $r_1 = q$, $f = \mathrm{ord}_p(q)$, and $d_1 = \left\lfloor \ln \frac{\mathrm{ord}_p(q)p}{2(p-1)} / \ln q \right\rfloor$ in Theorem 3.7. Then the assumptions are satisfied, for $t = q^{d_1} < fp/(2(p-1))$. Thus Theorem 2.4.10 gives the assertion. □

Example 2.4.12 We consider the classical example $p = 23$, $q = 2$. Corollary 3.8 shows that the order of a prime above 2 in the class group of $\mathbb{Q}(\xi_{23})$ is at least $\lfloor \ln \frac{11 \cdot 23}{2 \cdot 22} / \ln 2 \rfloor + 1 = 3$. Since the class number of $\mathbb{Q}(\xi_{23})$ is 3, such a prime generates the full class group.

It is straightforward to combine Corollary 2.4.11 with reciprocity laws to show that certain prime ideals are always nonprincipal. We only mention the following case containing the classical $p = 23$.

Corollary 2.4.13 *Let $p \geq 23$ be a prime $\equiv 7 \pmod 8$. Then the prime ideals above 2 in $\mathbb{Z}[\xi_p]$ are nonprincipal.*

Proof Since $p \geq 23$ and $\mathrm{ord}_{23}(2) = 11$, we have $\mathrm{ord}_p(2) \geq 5$. By quadratic reciprocity $\mathrm{ord}_p(2)$ is odd. Thus Corollary 2.4.11 shows that the order of a prime above 2 in the class group of $\mathbb{Q}(\xi_p)$ is at least $\lfloor (\ln 5/2)/\ln 2 \rfloor + 1 = 2$. □

2.4.2 Class groups of CM-fields

Now we turn our attention to the general case of CM-fields. We will use the idea of the "field descent" to obtain some explicit subgroups of class groups of CM-fields. In the next section, these results will be used to provide lower bounds on p-ranks of class groups.
For a group G and $g \in G$, we denote the order of g in G by $\mathrm{ord}(g)$. The ideal class group of a number field K is denoted by Cl_K. For a set L of ideals of a number field, we write $\overline{L} = \{\overline{I} : I \in L\}$ where the bar denotes the complex conjugation.

Theorem 2.4.14 *Let K be a CM-field, and let k be a complex subfield of K such that K/k is Galois. Let $\mathbb{Q}(\xi_m)$, $m \not\equiv 2 \pmod 4$, be the largest cyclotomic field contained in K.*
Let $D := \mathrm{Gal}(K/k)$, and $\{\sigma_1, ..., \sigma_s\}$ be a set of generators for D. Let $\mathcal{T}$ be a set of primes of k with $\mathcal{T} \cap \overline{\mathcal{T}} = \emptyset$. Denote the ramification index of $P \in \mathcal{T}$ in K/k by $R(P)$.
Then the ideal class group Cl_K of K contains a subgroup L isomorphic to Ω/Λ where

$$\Omega := \bigoplus_{P \in \mathcal{T}} (\mathbb{Z}/R(P)\mathbb{Z})$$

and Λ is a subgroup of Ω of rank at most s and exponent dividing $\lambda := \mathrm{lcm}(f(m, \sigma_i), i = 1, ..., s)$. Here f is the function defined in Theorem 2.1.4.

Moreover, L is contained in the subgroup of Cl_K generated by the primes of K above primes in $\mathcal{T}$.

Proof We first need some notation. Write $\mathcal{S} := \mathcal{T} \cup \overline{\mathcal{T}}$. Write $R(P) = R(\overline{P})$ for $P \in \overline{\mathcal{T}}$. We have $P = P_K^{R(P)}$ for $P \in \mathcal{S}$ where each P_K is an ideal of K. For the convenience of the reader, we give a table of the notations we need for this proof. All occurring ideals are viewed as ideals of K.

$K:$		CM-field
$K/k:$		Galois extension, k complex
$D:$	$=$	$\mathrm{Gal}(K/k)$
$\mathcal{T}:$	$=$	a set of primes of k with $\mathcal{T} \cap \overline{\mathcal{T}} = \emptyset$
$\mathcal{S}:$	$=$	$\mathcal{T} \cup \overline{\mathcal{T}}$
$R(P):$		the ramification index of $P \in \mathcal{S}$ in K/k
$P_K:$		the ideal of K with $P_K^{R(P)} = P$, $P \in \mathcal{S}$
$\Gamma:$		the group of all ideals of K
$H:$		the group of all principal ideals of K
$I:$	$=$	the subgroup of Γ generated by the ideals P_K, $P \in \mathcal{S}$
$I^*:$	$=$	the subgroup of I generated by the ideals P_K, $P \in \mathcal{T}$
$R:$	$=$	$\{J\overline{J} : J \in I\}$
$I_k:$	$=$	the subgroup of I generated by the ideals in $\mathcal{S}$
$K_\mathcal{S}:$	$=$	$\{Y/\overline{Y} : Y \in K, (Y/\overline{Y}) \in I\}$
$Z:$		the subgroup of I generated by $(Y_1), ..., (Y_s)$ (the Y_i are defined below)
$W:$	$=$	$\{J \in I : J/\overline{J} \in ZI_k\}$.

Claim 1: If $X \in K_\mathcal{S}$, then X^σ/X is a root of unity for all $\sigma \in D$.

Proof of Claim 1: Since σ fixes all ideals in I, we have $(X^\sigma) = (X)$. Thus $X^\sigma = \varepsilon X$ for some unit ε. But, since complex conjugation commutes with σ, we have $|\varepsilon|^2 = (X^\sigma/X)\overline{(X^\sigma/X)} = (|X|^2)^\sigma/|X|^2 = 1/1 = 1$. Thus ε is a root of unity by Lemma 2.4.1 c. This proves Claim 1.

Now we define some useful elements Y_i of K. Recall $D = \langle\sigma_1, ..., \sigma_s\rangle$. Define $F_1 := K_\mathcal{S}$ and $F_i := K_\mathcal{S} \cap \mathrm{Fix}\ \langle\sigma_1, ..., \sigma_{i-1}\rangle$ for $i = 2, ..., s+1$. For a root of unity ε, we write $\mathrm{ord}(\varepsilon)$ for the order of ε. Note that Y^{σ_i}/Y is a root of unity for all $Y \in F_i$ and $i = 1, ..., s$ by Claim 1. For $i = 1, ..., s$, let Y_i be an element of F_i such that

$$\mathrm{ord}(\frac{Y_i^{\sigma_i}}{Y_i}) = \max\{\mathrm{ord}(\frac{Y^{\sigma_i}}{Y}) : Y \in F_i\}.$$

Write $\eta_i := Y_i^{\sigma_i}/Y_i$. Note that

$$\operatorname{ord}(\frac{Y^{\sigma_i}}{Y}) \text{ divides } \operatorname{ord}(\eta_i) \tag{2.10}$$

for all i and all $Y \in F_i$. Furthermore, $\operatorname{ord}((Y\alpha_Y)^{\sigma_i}/(Y\alpha_Y))$ divides $\lambda = \operatorname{lcm}(f(m,\sigma_i), i = 1, ...s)$ for some root of unity α_Y for all $Y \in K_S$ by Theorem 2.1.4. Thus

$$(Y\alpha_Y)^\lambda \in k \quad \text{for all } Y \in K_S. \tag{2.11}$$

Claim 2: Let Z be the subgroup of I generated by $(Y_1), ..., (Y_s)$. If $X \in K_S$, then $(X) \in ZI_k$.

Proof of Claim 2: The assertion is proven if we can find an element Y of Z such that $XY \in k$. We recursively construct Y as follows. Define $X_1 := X$. Let $X_i \in F_i$, $1 \leq i \leq s$, be given. Then $X_i^{\sigma_i} = \eta_i^{j_i} X_i$ for some j_i by (2.10) and

$$X_{i+1} := X_i Y_i^{-j_i} \in \operatorname{Fix} \sigma_i$$

since $(Y_i^{-j_i})^{\sigma_i} = \eta_i^{-j_i} Y_i^{-j_i}$. As $X_i, Y_i \in F_i$, we get $X_{i+1} \in F_{i+1}$. So we have $X_{s+1} = X \prod_{i=1}^s Y_i^{-j_i} \in F_{s+1} = k$. Thus we may choose $Y = \prod_{i=1}^s Y_i^{-j_i}$ proving Claim 2.

Claim 3: We have $A \leq W$ for

$$\begin{aligned} A: &= I \cap I_k RH \\ W: &= \{J \in I : J/\overline{J} \in ZI_k\}. \end{aligned}$$

Proof of Claim 3: Let $X \in A$, say $X = i_k r(h)$ with $i_k \in I_k$, $r \in R$ and $h \in K$. Then $h/\overline{h} \in K_S$ and thus $(h/\overline{h}) \in ZI_k$ by Claim 2. Thus $X/\overline{X} = (h/\overline{h})(i_k/\overline{i_k}) \in ZI_k$ and hence $X \in W$ proving Claim 3.

Claim 4: Write $(Y_i) = A_i/\overline{A_i}$ with $A_i \in I^*$. Let Z^* be the subgroup of I^* generated by A_i, $i = 1, ...s$. Then

$$W = Z^* RI_k.$$

Proof of Claim 4: From the definitions, we have $Z^*RI_k \subset W$. It remains to show $W \subset Z^*RI_Q$. Let $X \in W$ be arbitrary. Write $X = a\overline{b}$ with $a, b \in I^*$. Then $X/\overline{X} = (a/b)(\overline{b}/\overline{a}) \in ZI_k$ and thus $a/b \in Z^*I_k$. Hence $X = a\overline{b} = (a/b)(b\overline{b}) \in Z^*I_kR$ concluding the proof of Claim 4.

Claim 5: The group W/RI_k has rank at most s and exponent at most λ.

Proof of Claim 5: The assertion on the rank follows from Claim 4 since W/RI_k is generated by $A_iRI_k : i = 1, ..., s$. Recall that $(Y_i) = A_i/\overline{A_i}$ and

$Y_i^\lambda \in k$ by (2.11). This implies $A_i^\lambda \in I_k$ and thus $\exp W/RI_k \leq \lambda$ proving Claim 5.

Claim 6: IRH/I_kRH has a factor group isomorphic to I/W.

Proof of Claim 6: Since $IRH/I_kRH \cong I/I \cap I_kRH$, the assertion follows from Claim 3.

Claim 7: We have

$$I/W \cong \left(\bigoplus_{P\in\mathcal{T}} (\mathbb{Z}/R(P)\mathbb{Z})\right)/U$$

for some subgroup U of $\bigoplus_{P\in\mathcal{T}}(\mathbb{Z}/R(P)\mathbb{Z})$ with rank $U \leq s$ and $\exp U \leq \lambda$.

Proof of Claim 7: This follows from $I/W \cong (I/RI_k)/(W/RI_k)$ and Claim 5 since $(I/RI_k) \cong \bigoplus_{P\in\mathcal{T}}(\mathbb{Z}/R(P)\mathbb{Z})$.

Conclusion of the proof of Theorem 2.4.14: From Claims 6 and 7 we know that IRH/I_kRH has a factor group and thus a subgroup isomorphic to

$$\left(\bigoplus_{P\in\sqcup} (\mathbb{Z}/R(P)\mathbb{Z})\right)/U.$$

This implies the assertion of the theorem. □

Corollary 2.4.15 *Let q be a prime. In the situation of Theorem 2.4.14, the following hold.*
a) The subgroup of Cl_K generated by the primes in $\mathcal{T}_K$ has order at least $[\prod_{P\in\mathcal{T}} R(P)]/\lambda^s$.
b) Let q be a prime, let d_q denote the q-rank of D, and let R_q be the number of $R(P)$s divisible by q. Then the q-rank of Cl_K is at least $R_q - d_q$.
c) If m and $|D|$ are odd and relative prime, then Cl_K contains a subgroup isomorphic to

$$\bigoplus_{P\in\mathcal{T}} (\mathbb{Z}/R(P)\mathbb{Z}).$$

Proof: Parts a and b follow directly from Theorem 2.4.14. For part c, note that, in this situation, $f(m,\sigma) = 1$ for all $\sigma \in D$ by Theorem 2.1.4. □

Note that in Theorem 2.4.14, the group D cannot contain the complex conjugation since it does not fix the occuring prime ideals. However, in the case where D consists only of the complex conjugation and the identity, we can prove the following analogon to Theorem 2.4.14. As we will explain later, we will end up in recovering a special case of a result of Schoof [119] which was obtained by class field theory and Galois cohomology.

Theorem 2.4.16 *Let K be a CM-field with maximal real subfield K^+ and ideal class group Cl_K. Let r be the number of finite primes ramified in K/K^+. Then*

$$d_2 Cl_K \geq r - 1.$$

Proof: Let I be the ideal group of K generated by the primes above the primes of K^+ ramified in K/K^+. Let $X \in K$ with $(X) \in I$. Then $\overline{X} = \varepsilon_X X$ for some unit ε since all ideals in I are invariant under complex conjugation. From Lemma 2.4.1 we know that ε_X is a root of unity since $|\varepsilon_X| = 1$. We choose an $X \in K$ such that the order of ε_X is maximum. Then, for every $Y \in K$ with $(Y) \in I$, there is a j such that $\overline{XY^j} = XY^j$ and thus $(XY^j) \in I^2$. Let H respectively H_I be the group of all principal ideals of K respectively all principal ideals in I. Then, by what we have seen, $H_I \leq I^2\langle(Y)\rangle$ and thus $I^2 H_I \leq I^2\langle(Y)$. Note that $I/I^2 \cong (\mathbb{Z}/2Z)^r$ and that $I^2\langle(Y)/I^2\rangle$ is of order at most 2. Thus $d_2 IH/I^2H = d_2 I/I^2 H_I \geq r - 1$. □

2.4.3 p-ranks and class fields towers

In this section, we use the fields descent method to obtain some lower bounds on p-ranks of ideal class groups of CM-fields. We also explain the connection of these results to infinite class field towers. There is a vast literature on these problems. One of the most useful lower bounds on p-ranks of ideal class groups is due to Schoof [119]. We will obtain a similar bound which only is applicable to CM-fields, but apperently is stronger in some cases.
The p-ranks of ideal class groups of number fields play an important role in algebraic number theory, see [18, 110, 119, 130], for instance. In particular, lower bounds on p-ranks of class groups are desirable. One of the reasons for the interest in these lower bounds is the connection to the problem of class field towers. We give a quick review of the basics on class field towers. Let K be an algebraic number field. Define $K_0 := K$ and for $n = 1, 2, ...$ let K_{n+1} be the Hilbert class field respectively the p-Hilbert class field of K_n for some prime p (see [18] for the terminology). Then

$$K_0 \subset K_1 \subset K_2 \subset \cdots$$

is called the **class field tower** respectively the **p-class field tower** of K. Such a class field tower is called **finite** if $\cup_{n=0}^{\infty} K_n$ is a finite extension of K and **infinite** otherwise. The existence of infinite class field towers had been conjectured for several decades and finally was proven by Golod and Shavarevic [49]. The following refinement of their result can be found in [18]. By E_K respectively Cl_K we denote the group of units respectively the ideal class group of an algebraic number field K. For a prime p and a group G, we denote the p-rank of G by $d_p G$.

Result 2.4.17 *Let K be an algebraic number field, and let p be a prime. If*

$$d_pCl_K \geq 2 + 2\sqrt{d_pE_K + 1},$$

then K has an infinite p-class tower and thus an infinite class field tower.

The value of d_pE_K can be determined by Dirichlet's unit theorem (see [13]). Thus the essential step for the application of Result 2.4.17 is the find lower bounds for d_pCl_K. Many results in this direction are known, see [18], for instance. We will compare results to one of the most useful lower bounds on d_pCl_K due to Schoof [119]. As mentioned in [119, p. 214], one can combine results from [19] with Proposition 3.1 of [119] to get the following.

Result 2.4.18 (Schoof) *Let K be an algebraic number field with ideal class group Cl_K and group of units E_K. Let p be a prime, and let K/k be a cyclic extension of degree p. Then*

$$d_pCl_K \geq \rho - d_pE_k - 1$$

where ρ is the number of finite and infinite primes of k ramified in K/k.

Schoof's Result [119] implies our Theorem 2.4.16 as can be seen as follows. Let K be a CM-field as in Theorem 2.4.16 and let $k = K^+$. The number of infinite primes of k ramified in K/k is $[k : \mathbb{Q}]$. Moreover, by Dirichlet's unit theorem (see [13]), we have $d_2E_k = [k : \mathbb{Q}]$. Thus Theorem 2.4.18 indeed shows that

$$d_2Cl_K \geq r + [k : \mathbb{Q}] - 1 - [k : \mathbb{Q}] = r - 1.$$

From Theorem 2.4.14, we get a bound similar to Schoof's result. Our result only holds for CM-fields, but as a compensation apperently sometimes gives stronger bounds.

Corollary 2.4.19 *Let K be a CM-field, and let k be a subfield of K such that K/k is Galois with Galois group D. Let p be a rational prime, and let ω be the number of finite primes of k which are not invariant under complex conjugation and whose ramification index in K/k is divisible by p. Then*

$$d_pCl_K \geq \frac{\omega}{2} - \varepsilon$$

where $\varepsilon = 1$ if K contains the pth roots of unity, and $\varepsilon = d_pD$ otherwise.

Proof This follows from Theorem 2.4.14 since $(f(m, \sigma), p) = 1$ for all $\sigma \in D$ if K does not contain the pth roots of unity. □

Corollary 2.4.20 *Let K be a CM-field, and let k be a subfield of K such that K/k is a Galois extension of prime degree p. Let ω be the number of primes of k ramified in K/k and not invariant under complex conjugation. Then*

$$d_pCl_K \geq \frac{\omega}{2} - \varepsilon$$

where $\varepsilon = 1$ if K contains the pth roots of unity and $\varepsilon = 0$ otherwise.

Often it is desirable to find number fields of low absolute degree whose class groups have large p-rank for some prime p. Therefore, the following result on cyclotomic fields is of interest.

Corollary 2.4.21 *Let $m = pm'$ where p is an odd prime with $(p, m') = 1$ such that -1 is not a power of p modulo m'. Let q be a prime divisor of $p-1$. Then the cyclotomic field $\mathbb{Q}(\xi_m)$ has a complex subfield K of absolute degree $\varphi(m')q/\mathrm{ord}_{m'}(p)$ with*

$$d_q Cl_K \geq \frac{\varphi(m')}{2\mathrm{ord}_{m'}(p)} - \varepsilon$$

where $\varepsilon = 1$ if $p = 2$ or q divides m' and $\varepsilon = 0$ otherwise.

Proof By Result 1.4.3, the decomposition group of the primes above p in $\mathbb{Q}(\xi_m)$ is generated by σ_p and $\sigma_{m'}$ where $\sigma_p(\xi_{p^a}) = \xi_p^x$ for a primitive root x modulo p, $\sigma_p(\xi_{m'}) = \xi_{m'}$, $\sigma_{m'}(\xi_p) = \xi_p$, and $\sigma_{m'}(\xi_{m'}) = \xi_{m'}^p$. Let K be the fixed field of $\langle\sigma_p^q, \sigma_{m'}\rangle$. Then K has absolute degree $q\varphi(m')/\mathrm{ord}_{m'}(p)$. There are $\varphi(m')/\mathrm{ord}_{m'}(\ )$ distinct primes of K above p which are all not invariant under complex conjugation by Result 1.4.3 since -1 is not a power of p modulo m'. The ramification index of these ideals in $K/\mathbb{Q}$ is q. Thus the assertion follows from Corollary 2.4.19 □

Corollary 2.4.22 *Let $m = pm'$ where p is an odd prime with $(p, m') = 1$ such that -1 is not a power of p modulo m'. Let q be a prime divisor of $p-1$. If*

$$\frac{\varphi(m')}{\mathrm{ord}_{m'}(p)} \geq 8q + 12, \tag{2.12}$$

then $\mathbb{Q}(\xi_m)$ has an infinite q-class field tower.
Furthermore, if q is odd and does not divide m', then the same conclusion still holds if (2.12) is replaced by

$$\frac{\varphi(m')}{\mathrm{ord}_{m'}(p)} \geq 8(q+1). \tag{2.13}$$

Proof Let K be the subfield of $\mathbb{Q}(\xi_m)$ defined in the proof of Corollary 2.4.21. Then, by Dirichlet's unit theorem, $d_q E_K \leq [\mathbb{Q} : K]/2 = \varphi(m')q/(2\mathrm{ord}_{m'}(p))$. Moreover, $d_q Cl_K \geq \varphi(m')/(2\mathrm{ord}_{m'}(p)) - 1$ by Corollary 2.4.21. Write $x := \varphi(m')/(2\mathrm{ord}_{m'}(p))$. By Result 2.4.17, K and thus $\mathbb{Q}(\xi_m)$ has an infinite q-class field tower if $x - 1 \geq 2 + 2\sqrt{qx+1}$, i.e., if $x \geq 4q + 6$. This proves the first assertion of Corollary 2.4.22. The second assertion follows by the same argument since in this case $d_q E_K \leq \varphi(m')q/(2\mathrm{ord}_{m'}(p)) - 1$ and $d_q Cl_K \geq \varphi(m')/(2\mathrm{ord}_{m'}(p))$. □

Corollary 2.4.23 *For every prime q, there are infinitely many cyclotomic fields $\mathbb{Q}(\xi_{rp})$, r, p prime, with infinite q-class field towers.*

Proof We use Dirichlet's theorem on primes in arithmetic progession (see [13]). We choose a prime $r \equiv 1 \pmod 3$ with $(r-1)/3 \geq 8q+12$. By Dirichlet's theorem, there are infinitely many primes p with $\mathrm{ord}_r(p) = 3$. By Corollary 2.4.22, $\mathbb{Q}(\xi_{rp}$ has an infinite class field tower for each of these p. □
The following bound on 2-ranks of subfields of cyclotomic fields is consequence of Theorem 2.4.16.

Corollary 2.4.24 *Let $m = m'p$ where p is an odd prime with $(p, m') = 1$. Let 2^b be the exact power of 2 dividing $p-1$. Let $\varepsilon = 0$ if -1 is a power of p modulo m' and $\varepsilon = 1$ otherwise. Then $\mathbb{Q}(\xi_m)$ has a complex subfield K of absolute degree $2^{b-\varepsilon}\varphi(m')/\mathrm{ord}_{m'}(p)$ with*

$$d_2 Cl_K \geq \varphi(m')/(2^\varepsilon \mathrm{ord}_{m'}(p)) - 1.$$

Proof Let G be the subgroup of $\mathrm{Gal}(\mathbb{Q}(\xi'_m)/\mathbb{Q})$ generated by $\sigma_p : \xi_{m'} \to \xi_{m'}^p$ together with the complex conjugation. Let E be the subfield of $\mathbb{Q}(\xi'_m)$ fixed by G, let F be the unique subfield of $\mathbb{Q}(\xi_p)$ of degree 2^b, and let $K := FG$. Then $[K : \mathbb{Q}] = 2^{b-\varepsilon}\varphi(m')/\mathrm{ord}_{m'}(p)$ as asserted. Let K^+ be the maximal real subfield of K. Then all $\varphi(m')/(2^\varepsilon \mathrm{ord}_{m'}(p))$ primes above p ramify in K/K^+. Thus the assertion follows from Theorem 2.4.16. □

Corollary 2.4.25 *Let $m = m'p$ where $p \equiv 3 \pmod 4$ is a prime with $(p, m') = 1$. Let $\varepsilon = 0$ if -1 is a power of p modulo m' and $\varepsilon = 1$ otherwise. If*

$$\frac{\varphi(m')}{2^\varepsilon \mathrm{ord}_{m'}(p)} \geq 10,$$

then $\mathbb{Q}(\xi_m)$ has an infinite 2-class field tower.

Proof This follows from Result 2.4.17 and Corollary 2.4.24. □

Remark 2.4.26 As mentioned in [119], the cyclotomic field with the smallest conductor known to have an infinite class field tower is $\mathbb{Q}(\xi_{363})$. Corollary 2.4.25 gives a new proof that this field has an infinite 2-class field tower. To see this, take $p = 3$, $m' = 121$. Then $\varepsilon = 1$, and $\varphi(m')/(2^\varepsilon \mathrm{ord}_{m'}(p)) = 110/(2 \cdot 5) = 11$.

Chapter 3

Exponent bounds

One of the central themes of the theory of difference sets is to search for necessary and sufficient conditions on the group structure for the existence of difference sets. We will focus on the necessary conditions here. Most known necessary conditions have the form of **exponent bounds**. Here the **exponent** of an abelian group means the order of its largest cyclic subgroup. An exponent bound gives an upper bound on the exponent of abelian groups containing difference sets. Turyn's exponent bound [127] from 1965 is the most prominent example. Since Turyn's work, exponent bounds have played an important role in the study of difference sets. In this chapter, we will derive new, more sophisticated exponent bounds. In the first part, we substantially improve Turyn's bound for several types of difference sets. This leads to necessary and sufficient conditions for the existence of difference sets in some cases. In the second part, we obtain a new, more general type of exponent bound by the method of the field descent.

3.1 Self-conjugacy exponent bounds

In 1965, Turyn [127] proved a bound on the exponent of abelian groups containing difference sets. Turyn's bound has two important features: It relies on a self-conjugacy assumption and is obtained by considering only a *single homomorphic image* of a putative difference set.

For the self-conjugacy assumption, see Remark 1.4.4 and Corollary 1.4.5. As indicated there, this assumption is very restrictive. Self-conjugacy is a happy but small world. Thus nonexistence results on difference sets without the self-conjugacy assumption are very desirable. However, only little progress has been made in this direction in the 35 years since Turyn's work. In Section 3.2, we will use the idea of the field descent to obtain some significant new exponent bounds not relying on the self-conjugacy assumption.

Now we discuss the second feature of Turyn's exponent bound, the use of just a single homomorphic image of a putative difference set. This means that only small part of the information on the difference set is used, and that there is hope for improvement. However, the combination of the information one gets from *more than one different* homomorphic images of a difference set usually is difficult. Results in this direction are very rare. In Section 3.1.2, we will obtain a powerful lemma for combining homomorphic images leading to dramatic improvements of Turyn's bounds for McFarland difference sets and relative difference sets with prime power parameters. In fact, these bounds are sharp. Further improvements of the Turyn bound for Chen and Davis-Jedwab difference sets are given in Sections 3.1.6 and 3.1.7

3.1.1 Turyn's exponent bound

Because of its importance and instructiveness we explain Turyn's exponent bound in detail before we come to our new results mentioned above. Turyn's bound depends on a self-conjugacy assumption and is proved by a "size argument" [129] which is best understood in terms of Ma's Lemma 1.5.1. Since the most general version of Turyn's bound is quite tedious to state, we confine ourselves to a simple version sufficing for our purposes.

Result 3.1.1 ([127]) *Let G be an abelian group of order v, and let n be a positive integer with $(v,n) > 1$. Let p be a prime divisor of (v,n) which is self-conjugate modulo* $\exp G$ *such that p^{2a} divides n for some $a \geq 1$. Assume the existence of $D \in \mathbb{Z}[G]$ with nonnegative coeffients bounded by a constant C such that*

$$DD^{(-1)} = n + AW \tag{3.1}$$

for some $A \in \mathbb{Z}[G]$ and some subgroup W of G of order divisible by p. Let G_p be the Sylow p-subgroup of G. Then

$$\frac{|G_p|}{\exp G_p} \geq \frac{p^a}{C}.$$

Proof Let W_p be the Sylow p-subgroup of W. Let U be a subgroup of G_p of order $G_p/\exp G_p$ not containing W_p such that G/U has a cyclic Sylow p-subgroup S. Let $\rho : G \to G/U$ be the canonical epimorphism. From (3.1), Lemma 1.3.3 and Corollary 1.4.5 we get $\chi(\rho(D)) \equiv 0 \pmod{p^a}$ for all characters χ of G/U of order divisible by $|S|$. Let P be the subgroup of G/U of order p. Then, by Ma's lemma 1.5.1, $\rho(D) = p^a X + PY$ for some $X, Y \in \mathbb{Z}[G/U]$ with nonnegative coefficients. Note $X \neq 0$ since otherwise $\rho(D)\rho(D)^{(-1)}$ is a multiple of P contracting (3.1) as $U \not\geq W$. Thus $\rho(D)$ has a coefficient $\geq p^a$. On the other hand, the coefficients of $\rho(D)$ are bounded by $C|G_p|/\exp G_p$. Hence $p^a \leq C|G_p|/\exp G_p$. □

Turyn's bound can be applied to difference sets, relative difference sets and group invariant weighing matrices. We only mention the application to difference sets which will be used in Section 3.2.3.

Corollary 3.1.2 ([127]) *Assume the existence of a (v, k, λ, n)-difference set in a group G. Let p be a prime divisor of v such that p^{2a} divides n for some $a \geq 1$. Let N be a normal subgroup of G such that G/N is abelian and p is self-conjugate modulo $\exp G/N$. Let P be the Sylow p-subgroup of G/N. Then*

$$\frac{|P|}{\exp P} \geq \frac{p^a}{|N|}.$$

3.1.2 The coset intersection lemma

We are going to prove a powerful lemma on "intersections" of group ring elements with cosets of subgroups. This result can be used to combine the information one gets on a group ring element through different homomorphic images. In Sections 3.1.3 and 3.1.4, we will apply this lemma to obtain sharp exponent bounds on McFarland difference sets and relative difference sets with prime power parameters. These bounds are substantial improvements of Turyn's exponent bounds.
Before the abstract formulation of our "intersection lemma", we try to draw a vivid picture of this important result by verbally explaining a simplified version. The situation we are interested in is as follows. Let $G = H \times W$ be an abelian group where W is a p-group. We are interested in subsets D of G which have a certain intersection pattern with the cosets of complements of W in G. More precisely, we are requiring that for each complement H' of W in G, there is exactly one coset of H' which has a large intersection with D, and all other cosets have a relatively small intersection with D. What we will show is that, under appropriate conditions, *exactly* the same coset intersection pattern inherits to all subgroups of complements of H in G. This result is strong and quite surprising since at first sight it seems "obvious" that something must be lost when switching to subgroups.

Now we come to the formulation of our "intersection lemma". Let G be a finite abelian group, and let P be be the Sylow p-subgroup of G. For any $a \in P$ and any subgroup $A = \langle b_1 \rangle \times \cdots \langle b_r \rangle$ of P such that $A \cap \langle a \rangle = \{1\}$ and $o(a) \geq \exp A$, define

$$\mathcal{S}(a, A) = \{U < P | \; U = \langle a_1 b_1 \rangle \times \cdots \langle a_r b_r \rangle, \; a_i \in \langle a \rangle, \; o(a_i) \leq o(b_i)\}.$$

Let $D = \sum_{g \in G} a_g g$ be an element of $\mathbb{Z}[G]$. For $X \subset G$, we define $D(X) := \sum_{g \in X} a_g$. Now we are ready to state the lemma.

Lemma 3.1.3 *Let $D = \sum_{g \in G} a_g g$ be an element of $\mathbb{Z}[G]$ with $a_g \geq 0$ for all g. Let $a \in P$, and let $A = \langle b \rangle \times W$ be a subgroup of P such that $A \cap \langle a \rangle = \{1\}$, $o(a) = p^t \geq \exp A$ and $o(b) \geq p$. Assume that there exists a positive integer δ, such that for any $U \in \mathcal{S}(a, A)$ and $g \in G$ either*

(1a) $D(Ug) - D(Uga^{p^{t-1}}) \geq \delta$ *and*
(1b) $D(Uga^{ip^{t-1}}) < \delta/p$ *for* $i = 1, ..., p-1$ *or*

(2) $D(Ug) < \delta/p$,

and there is at least one coset Ug *satisfying (1). Let* $B = \langle b^p \rangle \times W$. *Then for any* $U' \in S(a, B)$ *and* $g \in G$, *the coset* $U'g$ *satisfies either (1) or (2); and there is at least one coset* $U'g$ *satisfying (1).*

Proof We write $U' = \langle a_1 b^p \rangle \times V$ with $a_1 \in \langle a \rangle$, $o(a_1) \leq o(b^p)$ and $V \in S(a, W)$. Let $a_1' \in \langle a \rangle$, $(a_1')^p = a_1$. Define

$$U_i = \langle a_1' a^{ip^{t-1}} b \rangle \times V$$

for $i = 0, ..., p-1$. Note $U' < U_i$ and $U_i \in S(a, A)$. Let $g \in G$. If some $U_i g$ satisfies (2), then obviously U' also satisfies (2). Suppose that all $U_i g$ satisfy (1). We have $U_i = U' \sum_{j=0}^{p-1} (a_1')^j a^{ijp^{t-1}}$. Hence

$$\sum_{j=0}^{p-1} [D(U'(a_1')^j a^{ijp^{t-1}} g) - D(U'(a_1')^j a^{(ij+1)p^{t-1}} g)] \geq \delta$$

for $i = 0, ..., p-1$. Thus

$$\sum_{i=0}^{p-1} \sum_{j=0}^{p-1} [D(U'(a_1')^j a^{ijp^{t-1}} g) - D(U'(a_1')^j a^{(ij+1)p^{t-1}} g)] \geq p\delta.$$

If $j \neq 0$, then $\{a^{ijp^{t-1}} : i = 0, ..., p-1\} = \{a^{(ij+1)p^{t-1}} : i = 0, ..., p-1\}$. Hence

$$D(U'g) - D(U'ga^{p^{t-1}}) \geq \delta,$$

i.e. the coset $U'g$ satisfies (1a). It is clear that $U'g$ also satisfies (1b).
It remains to show that at least one coset $U'g$ satisfies (1). It is given that there is a coset $U_0 g$ satisfying (1). Hence $D(\langle a^{p^{t-1}} \rangle \times \langle a_1' b \rangle \times Vg) \geq \delta$. As

$$\langle a^{p^{t-1}} \rangle \times \langle a_1' b \rangle \times Vg = \bigcup_{j=0}^{p-1} U_i a^{jp^{t-1}} g$$

for $i = 0, ..., p-1$ there must be $j_i \in \{0, ..., p-1\}$, such that the coset $U_i g a^{j_i p^{t-1}}$ satisfies (1) ($i = 0, ..., p-1$). Since $U_0 g a^{j_0 p^{t-1}} \cap U_1 g a^{j_1 p^{t-1}} \neq \emptyset$, we can assume $j_0 = j_1 = 0$. As $U' = U_0 \setminus \bigcup_{j=1}^{p-1} U_1 a^{jp^{t-1}}$ and $D(U_1 g a^{jp^{t-1}}) < \delta/p$ for $j = 1, ..., p-1$, it follows that $D(U'g) > \delta/p$. Thus $U'g$ must be contained in $U_i g a^{j_i p^{t-1}}$ for all i, i.e. $j_i = 0$ for all i. By the same argument as above, it follows that the coset $U'g$ satisfies (1a) and (1b). □

Corollary 3.1.4 *In the situation of the Lemma 2.1 we have*

$$\delta \leq \max\{a_g : g \in G\}.$$

Proof Apply Lemma 2.1 repeatedly until $A = \{1\}$. □

3.1.3 McFarland difference sets

A **McFarland difference set** is a difference set with parameters

$$\begin{aligned} v &= q^{d+1}[1+(q^{d+1}-1)/(q-1)], \\ k &= q^d(q^{d+1}-1)/(q-1), \\ \lambda &= q^d(q^d-1)/(q-1), \\ n &= q^{2d} \end{aligned} \tag{3.2}$$

where $q = p^f \neq 2$ and p is a prime. We call (3.2) **McFarland parameters** determined by the triple (p, f, d). The case $q = 2$ is excluded since then $(v, k, \lambda) = (2^{2d+2}, 2^{2d+1} - 2^d, 2^{2d} - 2^d)$, i.e., we are dealing with *Hadamard* difference sets in 2-groups. The following result is due to McFarland [95].

Result 3.1.5 *Let G be an abelian group of oder $q^{d+1}[1+(q^{d+1}-1)/(q-1)]$ where $q = p^f$ and p is a prime. Let P denote the Sylow p-subgroup of G. If p is odd and* $\exp P = p$ *or if $p = 2$ and $P \cong (\mathbb{Z}/2\mathbb{Z})^{fd+1}$ or $P \cong (\mathbb{Z}/2\mathbb{Z})^{fd-1} \times (\mathbb{Z}/4\mathbb{Z})$, then G contains a McFarland difference set.*

It is natural to ask wether there are further groups containing McFarland difference sets. Dillon [39] showed that McFarland's construction can be extended to certain nonabelian groups. An extension to other abelian groups of low exponent was obtained by Davis and Jedwab [33]. In the present section, we will obtain sharp exponent bounds on the Sylow p-subgroups of abelian groups containing McFarland difference sets under the self-conjugacy assumption. In several cases, these conditons are necessary and sufficient for the existence of McFarland difference sets. Parts of our proof depend on results of my doctoral dissertation [113] which will not be repeated here. These results also can be found in [86], equations (3.3), (4.3) and (4.4).

Theorem 3.1.6 *Assume that there is a McFarland difference set D in an abelian group G of order $q^{d+1}[1+(q^{d+1}-1)/(q-1)]$ where $q = p^f$ and p is a prime self-conjugate modulo* $\exp G$. *Let P be the Sylow p-subgroup of G. Then the following hold.*
(a) If p is odd, then P is elementary abelian.
(b) If $p = 2$ and $f \geq 2$, then $\exp P \leq 4$.

Proof Let a be an element of P order $p^e := \exp P$ and write $P = \langle a\rangle \times A$ for a suitable $A < P$. We use the notation of Lemma 3.1.2. Note that $\mathcal{S}(a, A)$ is the set of all complements of $\langle a\rangle$ in P. Let $U \in \mathcal{S}(a, A)$, and let $\rho : G \to G/U$

denote the canonical epimorphism. For $b \geq 0$ let P_b be the subgroup of G/U of order p^b.
a) Asssume that p is odd and $e \geq 2$. By [113, (3.3)] or [86, (3.3)] we have

$$\rho(D) = p^{fd}h + p^{fd-e}P_e B$$

where $h \in G/U$ and $B \subset G/U$ such that not two elements of $B \cup \{h\}$ are in the same coset of P_e in G/U. Thus the assumptions of Lemma 3.1.2 are satisfied with $\delta = p^{fd}$. But this contradicts Corollary 3.1.4 proving part a of the theorem.
b) Asssume $p = 2$ and $e \geq 3$. By [86, (4.3), (4.4)] or [113, (3.7), (3.9)], we have

$$\rho(D) = 2^{fd}h + 2^{fd-e+1}P_{e-1}B + 2^{fd-e}P_e C$$

with $h \in G/U$ and $B, C \subset G/U$ such that no two elements of $B \cup \{h\}$ are in the same coset of P_{e-1} in G/U. Thus the assumptions of Lemma 3.1.2 are satisfied with $\delta = 2^{fd} - 2^{fd-e}$. But this contradicts Corollary 3.1.4. □

Corollary 3.1.7 *Let G be an abelian group G of order $q^{d+1}[1 + (q^{d+1} - 1)/(q-1)]$ where $q = p^f$ and p is a prime self-conjugate modulo $\exp G$. Let P be the Sylow p-subgroup of G. If p is odd, then a McFarland difference set in G exists if and only if $\exp P = p$. If $p = f = 2$, then a McFarland difference set in G exists if and only if $\exp P \leq 4$.*

Proof For odd p, the assertion follows from Result 3.1.5 and Theorem 3.1.6 a. For $p = f = 2$, it was shown by Davis and Jedwab [33] that G contains a McFarland difference set if $\exp P \leq 4$. By Theorem 3.1.6 b, this condition also is necessary for the existence of a McFarland difference set in G. □

3.1.4 Semiregular relative difference sets

In this section, we study relative difference sets with parameters $(m, n, k, \lambda) = (p^a, p^b, p^a, p^{a-b})$ where p is a prime. These difference sets are closely connected to finite affine and projective planes since $(p^a, p^a, p^a, 1)$ relative difference sets are equivalent to affine planes admitting a certain point regular automorphism group, see Proposition 1.2.4. The corresponding projective planes are called **quasiregular**, see [38]. The problem to find new exponent bounds for (p^a, p^b, p^a, p^{a-b}) relative difference sets, in particular, for even a, was raised by Pott [105, p. 109]. In order to understand the urgent need for a new exponent bound in the case that a is even, say $a = 2c$, a comparison with the case where a is odd, say $a = 2d + 1$, is enlightening. We will only discuss the case where p is odd here. The case $p = 2$ is quite different, see [105, 113]. Ma and Pott [84] proved that the exponent of an abelian group of order p^{2d+b+1} containing a $(p^{2d+1}, p^b, p^{2d+1}, p^{2d-b+1})$ relative difference set cannot exceed p^{d+1}. This exponent bound is satisfactory, as it is known from a construction of Davis [32] that it can be attained for all d and all $b \leq d$.

For relative $(p^{2c}, p^b, p^{2c}, p^{2c-b})$ difference sets, the situation is not that nice. There are several rather strong nonexistence results, see [104, 114], but no exponent bound comparable to the one for $(p^{2d+1}, p^b, p^{2d+1}, p^{2d-b+1})$-RDSs had been known. We will close this gap here by showing that for an odd prime p an abelian group containing a $(p^{2c}, p^b, p^{2c}, p^{2c-b})$-RDS cannot have an exponent exceeding p^{c+1}. This bound is satisfactory because it is known that it can be attained for all c and all $b \leq c$, see [32] or [105]. Together with the previosly known results, we obtain a unified sharp exponent bound for all relative (p^a, p^b, p^a, p^{a-b}) with p odd. The proof of our result is much more involved than those for the previously known exponent bounds. It relies on our "intersection lemma" 3.1.2 and a careful analysis of homomorpic images of putative relative difference sets.
Before we state our theorem, we recall an exponent bound on the forbidden subgroup due to Ma and Pott [84]. It will help us avoiding an unwelcome case distinction in the proof of Theorem 3.1.9.

Result 3.1.8 *Let G be an abelian group of order p^{2a+b}, and let N be a subgroup of G of order p^b. If there is a $(p^{2a}, p^b, p^{2a}, p^{2a-b})$ difference set in G relative to N, then*

$$\exp N \leq p^a.$$

Now we are ready to state the unified exponent bound.

Theorem 3.1.9 *Let p be an odd prime. If an abelian group G of order p^{a+b} contains a relative (p^a, p^b, p^a, p^{a-b}) difference set, then $\exp G \leq p^{\lfloor a/2 \rfloor + 1}$.*

Proof For odd a, a proof can be found in [84]. Hence we only need to consider even a, say $a = 2c$. Let R denote the putative relative difference set and assume $\exp G \geq p^{c+2}$, say $\exp G = p^t = p^{c+r+2}$ with $r \geq 0$. We show that this assumption leads to a contradiction. Write $G = \langle g \rangle \times H$ where g is an element of G of order p^t. We will show that the assumptions of Lemma 3.1.2 are satisfied for $P = G$, $A = H$, $D = R$, $h = g$ and $\delta = p^c - 2p^{c-r-2}$. Then Corollary 3.1.4 will imply $\delta \leq 1$ which is the desired contradiction. We note that, using notation of Lemma 3.1.2, $S(g, H)$ is the set of all complements of $\langle g \rangle$ in G. Let U be any of these complements. Then G/U is cyclic of order p^t.
First of all, we will show that the forbidden subgroup N cannot be contained in U. Assume the contrary. We choose a character χ of G with $Ker\chi \cap \langle g \rangle = \{1\}$ and $|Ker\chi \cap N| = |N|/p$. Write $K = Ker\chi$ and let $\tau : G \to G/K$ denote the canonical epimorphism. Since no two elements of R are in the same coset of N, the coefficients of $\tau(R)$ cannot exceed $|K|/|K \cap N| \leq p^{c-r-1}$. However, we know from Lemma 1.3.6 and Ma's lemma 1.5.1 that $\tau(R) = p^c X_1 + P' X_2$ where P' is the subgroup of order p of G/K and X_1, X_2 are elements of the group ring $\mathbb{Z}[G/K]$ with nonnegative coefficients. If we also view χ as a character of G/K, we obtain $\chi(R) = p^c \chi(X_1)$, and this implies $X_1 \neq 0$,

since $\chi(R) \neq 0$ by Lemma 1.3.6. Hence $\tau(R)$ has a coefficient $\geq p^c$. This contradicts the upper bound for the coefficients of $\tau(R)$ obtained above and shows that N indeed cannot be contained in U.
Now let $\rho : G \to G/U$ denote the canonical epimorphism. By the same argument as above, we see that the coefficients of $\rho(R)$ are bounded by $|U|/|U \cap N|$ and that $\rho(R)$ has at least one coefficient $\geq p^c$ (note that the argument for the existence of a coefficient $\geq p^c$ requires that N is not contained in U). Hence $p^c \leq |U|/|U \cap N|$ and

$$|\rho(N)| = \frac{|N|}{|U \cap N|} \geq \frac{p^c|N|}{|U|} = p^{r+2}.$$

We write $|\rho(N)| = p^x$ with $x \geq r+2$. By Result 3.1.8, we can assume $x \leq c$. From Lemmas 1.3.6, 1.5.3 and Corollary 1.4.5, we get (using the notation of Lemma 1.5.3)

$$\rho(R) = \sum_{m=0}^{x-1} \epsilon_m p^{c-m-1}(pP_m - P_{m+1})g_m + P_x Y.$$

Since $\psi(\rho(R)) = 0$ and $\psi(pP_m - P_{m+1}) = 0$ for $m = 0, ..., x-1$ for all $\psi \in P_x^\perp$ by Lemma 1.3.6, we conclude $\varphi(P_x Y) = 0$ for all nontrivial characters φ of G/U. By the Fourier inversion formula 1.3.4, $P_x Y$ must be a multiple of $G/U = P_{c+r+2}$. As $|R| = p^{2c}$, we must have $P_x Y = p^{c-r-2}P_{c+r+2}$; thus

$$\rho(R) = \sum_{m=0}^{x-1} \epsilon_m p^{c-m-1}(pP_m - P_{m+1})g_m + p^{c-r-2}P_{c+r+2}. \tag{3.3}$$

We claim

$$\epsilon_0 = \epsilon_1 = \cdots = \epsilon_{r+1} = 1 \text{ and } P_i g_0 = P_i g_i \tag{3.4}$$

for $i = 0, 1, ..., r+1$.
We prove (3.4) by induction. For $h \in G/U$ let $C(h)$ be the coefficient of g in $\rho(R)$.
(a) Assume $\epsilon_0 = -1$. Then by (3.3) (recall that $p > 2$)

$$\begin{aligned} C(g_0) &\leq -p^c + p^{c-1} + p^{c-1} - p^{c-2} + p^{c-2} + - \cdots + p^{c-x+1} - p^{c-x} + p^{c-r-2} \\ &= -p^c + 2p^{c-1} - p^{c-x} + p^{c-r-2} < 0, \end{aligned}$$

a contradiction. Hence $\epsilon_0 = 1$.
(b) Let $1 \leq l \leq r+1$, $\epsilon_0 = \epsilon_1 = \cdots = \epsilon_{l-1} = 1$ and $P_i g_0 = P_i g_i$ for $i = 0, 1, ..., l-1$. We have to show $\epsilon_l = 1$ and $P_l g_0 = P_l g_l$. From (3.3) we have

$$\rho(R) = (p^c - p^{c-l}P_l)g_0 + \sum_{m=l}^{x-1} \epsilon_m p^{c-m-1}(pP_m - P_{m+1})g_m + p^{c-r-2}P_{c+r+2}.$$

Let $g' \in P_l g_0 \setminus \{g_0\}$. If $\epsilon_l = -1$ or $P_l g_0 \neq P_l g_l$, then

$$\begin{aligned} C(g') &\leq -p^{c-l} + p^{c-l-1} + p^{c-l-1} - p^{c-l-2} + - \cdots + p^{c-x+1} - p^{c-x} + p^{c-r-2} \\ &= -p^{c-l} + 2p^{c-l-1} - p^{c-x} + p^{c-r-2} < 0, \end{aligned}$$

a contradiction. Thus we have proved (3.4). Hence we get

$$\rho(R) = (p^c - p^{c-r-2}P_{r+2})g_0 + \sum_{m=r+2}^{x-1} \epsilon_m p^{c-m-1}(pP_m - P_{m+1})g_m + p^{c-r-2}P_{c+r+2}$$

from (3.3). We infer

$$\begin{aligned} C(g_o) &\geq p^c - p^{c-r-2} + p^{c-r-3} - p^{c-r-3} + - \cdots - p^{c-x+1} + p^{c-x} \\ &= p^c - p^{c-r-2} + p^{c-x}, \\ C(h) &\leq -p^{c-r-2} + p^{c-r-2} - p^{c-r-3} + - \cdots + p^{c-x+1} - p^{c-x} + p^{c-r-2} \\ &= p^{c-r-2} - p^{c-x} \end{aligned}$$

for $h \in P_{r+2}g_0 \setminus \{g_0\}$ and

$$\begin{aligned} C(h') &\leq p^{c-r-2} - p^{c-r-3} + p^{c-r-3} - + \cdots + p^{c-x+1} - p^{c-x} + p^{c-r-2} \\ &= 2p^{c-r-2} - p^{c-x} \end{aligned}$$

for $h' \in (G/U) \setminus P_{r+2}g_0$. As $\rho(R)$ has at least one coefficient $\geq p^c$ we get $C(g_0) \geq p^c$.
Together with the upper bounds on $C(h)$ and $C(h')$ obtained above, this shows that U satisfies the conditions of Lemma 3.1.2 with $\delta = p^c - 2p^{c-r-2}$. Since U was chosen as an arbitrary element of $S(g, H)$, we indeed have indeed verified that Lemma 3.1.2 can be applied, and this proves the theorem. □

3.1.5 Two recent families of difference sets

Quite recently, two new families of difference sets were discovered by Chen [22] and Davis, Jedwab [33]. Chen's difference sets have parameters

$$\begin{aligned} v &= 4q^{2t}\frac{q^{2t}-1}{q^2-1}, \\ k &= q^{2t-1}[\frac{2(q^{2t}-1)}{q+1}+1], \\ \lambda &= q^{2t-1}(q-1)\frac{q^{2t-1}+1}{q+1}, \\ n &= q^{4t-2} \end{aligned}$$

where $q = p^f$ is a power of 3 or a square of an odd prime power and t is any positive integer. For $t = 1$, such a difference set is a Hadamard difference set. For $t \geq 2$, *any* difference set with the above parameters, for *any* prime power q, will be called a **Chen difference set**. Chen's construction only works for groups having an elementary abelian Sylow p-subgroup. It is natural to ask if Chen's construction can be extended to other groups. In some cases, we will be able to show that groups with Sylow p-subgroups of higher exponent cannot contain difference sets. Thus we obtain a necessary and sufficient condition for the existence of Chen difference sets in abelian groups in these cases. However, in general there remains a large gap between the known necessary and sufficient conditions on the existence of Chen difference sets. I do not know if the exponent bound obtained here can be improved.
The second recent series of difference sets was constructed by Davis and Jedwab [33] and has parameters

$$\begin{aligned} v &= 2^{2t+2}(2^{2t}-1)/3, \\ k &= 2^{2t-1}(2^{2t+1}+1)/3, \\ \lambda &= 2^{2t-1}(2^{2t-1}+1)/3, \\ n &= 2^{4t-2} \end{aligned}$$

where $t \geq 2$ is a positive integer. *Any* difference set with such parameters will be called a **Davis-Jedwab difference set**. Note that Davis-Jedwab difference sets are also Chen difference sets (put $q = 2$). Davis and Jedwab [33] constructed Davis-Jedwab difference sets in all abelian groups of order $2^{2t+2}(2^{2t}-1)/3$ which have a Sylow 2-subgroup S_2 of exponent at most 4, with the single exception of $t = 2$ and $S_2 \cong \mathbb{Z}_4^3$. This exception was removed by Arasu and Chen [1] who constructed the necessary difference set in $\mathbb{Z}_4^3 \times \mathbb{Z}_5$. We will show here that the condition $\exp S_2 \leq 4$ is also *necessary* for the existence of abelian Davis-Jedwab difference sets under the self-conjugacy assumption.
The following consequence of the Fourier inversion formula will be needed in the proofs of Theorems 3.1.12 and 3.1.15.

Lemma 3.1.10 *Let G be a finite abelian group, and let t be a positive integer. If $B \in \mathbb{Z}[G]$ with*

$$\chi(B) \equiv 0 \bmod t$$

for all characters of G then

$$B \equiv 0 \bmod t/(|G|, t).$$

3.1.6 Chen difference sets

In this section, we will improve Turyn's exponent bound in the case of Chen difference sets with odd q. In some cases, this will give a necessary and

sufficient condition for the existence of these difference sets. We will use arguments similar to those of Arasu, Davis and Jedwab [2]. A very nice description of this method can be found in [105].
Before we come to our result, we state what Turyn's exponent bound 3.1.2 yields for Chen difference sets.

Proposition 3.1.11 *Let $q = p^f$ be a prime power, and let G be an abelian group of order $4q^{2t}(q^{2t}-1)/(q^2-1)$ containing a Chen difference set. Assume that p is self-conjugate modulo* $\exp G$. *Denote the Sylow p-subgroup of G by S_p. Then the following hold.*
a) If p is odd, then $\exp S_p \leq q$.
b) If $p = 2$, then $\exp S_2 \leq 4q$.

Proof a) Putting $N = \{1\}$ and $a = f(2t-1)$ in Corollary 3.1.2 we get $|S_p|/\exp S_p \geq q^{2t-1}$. This implies the assertion since $|S_p| = q^{2t}$.
b) Again, take $N = \{1\}$ and $a = f(2t-1)$ in Corollary 3.1.2. This yields the assertion since $|S_2| = 4q^{2t}$. □

The following result shows that equality cannot hold in Turyn's exponent bound for Chen difference sets with odd p.

Theorem 3.1.12 *Let $q = p^f$ be an odd prime power, and let $t, f \geq 2$ be integers. Let G be an abelian group of order $4q^{2t}(q^{2t}-1)/(q^2-1)$ containing a Chen difference set D. Assume that p is self-conjugate modulo* $\exp G$. *Denote the Sylow p-subgroup of G by S_p. Then* $\exp S_p \leq p^{f-1}$.

Proof By Proposition 3.1.11, we have $\exp S_p \leq q$. Thus it suffices to show that $\exp S_p = q$ is impossible. To this end, we assume $S_p = \mathbb{Z}_q \times H$ where H is a subgroup of S_p of order q^{2t-1} and exponent at most q. Let K be any complement of $\mathbb{Z}_q$ in S_p, and let $\rho : G \to G/K$ be the canonical epimorphism. By Ma's lemma 1.5.1, we have

$$\rho(D) = q^{2t-1}X + PY \tag{3.5}$$

where X, Y are elements of $\mathbb{Z}[G/K]$ having nonnegative coefficients, and P is the subgroup of G/K of order p. Since $|K| = q^{2t-1}$, the coefficients of $\rho(D)$ cannot exceed q^{2t-1}. Thus X has coefficients 0 and 1 only, i.e., X is a subset of G/K. We can assume that no coset Pg of P is completely contained in X since otherwise we can replace Y by $Y + q^{2t-1}g$ and thus remove Pg from X. We write $w_g = |X \cap Pg|$ for $g \in G/K$ and $Y = \sum_{g \in T} a_g g$ where T is a set of distinct coset representatives of P in G/K. In view of (3.5), the coefficient of 1 in $\rho(D)\rho(D)^{(-1)}$ is $p \sum a_g^2 + q^{4t-2} \sum w_g$. Hence Lemma 1.3.2 gives us

$$p \sum a_g^2 + q^{4t-2} \sum w_g = q^{4t-2} + q^{2t-1}\lambda. \tag{3.6}$$

Let L be the preimage of P under ρ, and let $\psi : G \to G/L$ be the canonical epimorphism. From (3.5) we see that the coefficient of 1 in $\psi(D)\psi(D)^{(-1)}$ is $p^2 \sum a_g^2 + q^{4t-2} \sum w_g^2$. Hence Lemma 1.3.2 implies

$$p^2 \sum a_g^2 + q^{4t-2} \sum w_g^2 = q^{4t-2} + pq^{2t-1}\lambda. \tag{3.7}$$

From (3.6) and (3.7) we infer

$$\sum_{g \in T} (pw_g - w_g^2) = p - 1. \tag{3.8}$$

Since $0 \leq w_g \leq p-1$, we conclude that $w_h = 1$ or $p-1$ for exactly one h and $w_g = 0$ for all $g \neq h$.

Now, fix any complement of $\mathbb{Z}_q$ in S_p, say H. Let $\alpha : G \to G/H$ be the canonical epimorphism. From Lemmas 1.3.7, 3.1.10 and Corollary 1.4.5 we know that $\alpha(D)$ is divible by q^{2t-2}, say $\alpha(D) = q^{2t-2}u = \sum_{g \in G/H} b_g g$. Then by Lemma 1.3.2,

$$uu^{(-1)} = q^2 + q^2(q-1)\frac{q^{2t-1}+1}{q+1}(G/H).$$

Hence $\sum b_g^2 = q^2 + q^2(q-1)\frac{q^{2t-1}+1}{q+1}$. Furthermore, $\sum b_g = q(\frac{2(q^{2t}-1)}{q+1} + 1)$ and $|G/H| = 4q\frac{(q^{2t}-1)}{q^2-1}$.

Define $c_g := b_g - (q-1)/2$. Then we have the useful formula

$$\sum_{g \in G/H} c_g^2 = q^2 \tag{3.9}$$

whose proof is straighforward by using the expressions for $\sum b_g$, $\sum b_g^2$ and $|G/H|$ we obtained.

Since $\exp H \leq q$ and $t \geq 2$, the rank of H must be at least three. Let $g_1, ..., g_r$, $r \geq 3$, be a basis of H, and let

$$K_{ijk} = \langle g_1 z^i, g_2 z^j, g_3 z^k, g_4, ..., g_r \rangle$$

for $i, j, k = 0, ..., p-1$, where z is an element of order p of $\mathbb{Z}_q$. Then each K_{ijk} is a complement of $\mathbb{Z}_q$ in S_p and $K_{ijk}K_{i'j'k'} = H\langle z \rangle$ for all $(i,j,k) \neq (i',j',k')$. From the conclusion following (3.8), we know that for every triple (i,j,k), there is a coset L_{ijk} of $K_{ijk}\langle z \rangle = H\langle z \rangle$ such that either

(i) there is a coset of K_{ijk} in L_{ijk} which is completely contained in D and all other cosets of K_{ijk} in L_{ijk} have an empty intersection with D or
(ii) there are $p-1$ cosets of K_{ijk} in L_{ijk} which are completely contained in D and the remaining coset has an empty intersection with D.

We conclude $L_{ijk} \neq L_{i'j'k'}$ for $(i,j,k) \neq (i',j',k')$. Otherwise $L_{ijk} = L_{i'j'k'}$ would have to be contained in D since every coset of K_{ijk} in L_{ijk} meets every coset of $K_{i'j'k'}$ in L_{ijk}.
Furthermore, we observe that every coset L_{ijk}, $(i,j,k) \neq (0,0,0)$, corresponds to p coefficients $b_g = p^{f-1}$ or $b_g = (p-1)p^{f-1}$ in $u = \alpha(D)/q^{2t-2}$.
If $b_g = p^{f-1}$ then $c_g = \frac{1}{2}(-(p-2)p^{f-1}+1)$, and if $b_g = (p-1)p^{f-1}$ then $c_g = \frac{1}{2}((p-2)p^{f-1}+1)$. In both cases, we have $|c_g| \geq \frac{1}{2}((p-2)p^{f-1}-1)$.
We also know that $b_g = q$ for at least one g since there is a $w_g \geq 1$. As there are (p^3-1) cosets L_{ijk}, $(i,j,k) \neq (0,0,0)$, it follows that

$$4 \sum_{g \in G/H} c_g^2 \geq (q+1)^2 + (p^3-1)p[(p-2)p^{f-1}-1]^2. \tag{3.10}$$

For $p=3$ and $f=2$, we get $4\sum_{g\in G/H} c_g^2 \geq 100 + 26 \cdot 3 \cdot 4 = 412$ which yields a contradiction since $4\sum_{g\in G/H} c_g^2 = 4q^2 = 324$ by (3.9). For $(p,f) \neq (3,2)$ we get

$$\begin{aligned} 4 \sum_{g \in G/H} c_g^2 &> (p^3-1)p(p^{f-1}-1)^2 \\ &> (p^3-1)q(p^{f-1}-2) \\ &= q(p^{f+2}-p^{f-1}-2p^3+2) \\ &= q^2(p^2-1/p-2p^{3-f}+2/p^f) \\ &> 4q^2 \end{aligned}$$

(the last step uses $(p,f) \neq (3,2)$). This again contradicts (3.9). □

Corollary 3.1.13 *Let $q = p^f$ be an odd prime power, and let G be an abelian group of order $4q^4(q^2+1)$. If the Sylow 2-subgroup S_2 is isomorphic to $\mathbb{Z}_2^3$ and a Chen difference set exists in G, then the exponent of the Sylow p-subgroup S_p of G is at most p^{f-1}. In particular, if $f=2$ and $S_2 \cong \mathbb{Z}_2^3$, then a Chen difference set in G exists if and only if S_p is elementary abelian.*

Proof If $S_2 \cong \mathbb{Z}_2^3$, then $\exp G$ divides $q^4(q^2+1)$. Thus p is self-conjugate modulo $\exp G$, and the assertion follows from Corollary 1.4.5 and Theorem 3.1.12. □

Remark 3.1.14
a) The estimates used in the proof of Theorem 3.1.12 are rather crude. However, it is not clear if this can be viewed as evidence for a possible improvement of the exponent bound.
b) The method used in the proof of Theorem 3.1.12 does not work for $p=2$. In the next section, we present a method for $p=2$, $f=1$. It remains an open question if the exponent bound in Proposition 3.1.11 b can be attained for $p=2$, $f>1$.

3.1.7 Davis-Jedwab difference sets

As mentioned above, Davis-Jedwab difference sets exist in in all abelian groups G of order $2^{2t+2}(2^{2t}-1)/3$, $t \geq 2$, which have a Sylow 2-subgroup S_2 of exponent at most 4. All these difference sets have the property that the squareroot of the order of the difference set D divides all character values of D. We call this the **character divisibility property**. In the case $t = 2$, *all* putative Davis-Jedwab difference have this property by Corollary 1.4.5 since then 2 is self-conjugate modulo $|G|$. Note that 2 is not self-conjugate modulo $\exp G$ for any abelian group G of order $2^{2t+2}(2^{2t}-1)/3$, $t \geq 3$. Thus the self-conjugacy argument does not work in this case. However, it is still interesting to classify Davis-Jedwab difference sets having the character divisibility property. This is done in the following.

Theorem 3.1.15 *Let G be an abelian group of order $2^{2t+2}(2^{2t}-1)/3$, $t \geq 2$, with Sylow 2-subgroup S_2. A Davis-Jedwab difference set D in G having the character divisibility property exists if and only if $\exp S_2 \leq 4$.*

Proof

By the remarks above, we only need to prove the nonexistence part. Note that Corollary 3.1.2 remains true if the self-conjugacy assumption is replaced by the character divisibility property. Putting $N = 1$ and $a = 2^{2t-1}$ there, we conclude $\exp S_2 \leq 8$. It remains to show $\exp S_2 \neq 8$. Assume the contrary and write $S_2 = \mathbb{Z}_8 \times H$, where $\mathbb{Z}_8$ is cyclic of order 8 and $\exp H \leq 8$. If $\operatorname{rank} H = 1$ then $t = 2$ and $S_2 \cong \mathbb{Z}_8^2$. It is shown in [5] that this case cannot occur. Hence we can assume $\operatorname{rank} H \geq 2$.

Let U be any be any complement of $\mathbb{Z}_8$ in S_2, and let $\rho : G \to G/U$ be the canonical epimorphism. From Lemma 3.1.10 we get $\rho(D) \equiv 0 \pmod{2^{2t-4}}$, say $\rho(D) = 2^{2t-4}w_U$, $w_U = \sum_{g \in G/U} a_g g$. Ma's lemma 1.5.1 gives

$$w_U = 8X + PY \tag{3.11}$$

where P is the subgroup of order 2 in G/U and X, Y are elements of $\mathbb{Z}[G/U]$ with nonnegative coefficients. Since $\rho(D)$ cannot have coefficients greater than $|U| = 2^{2t-1}$, we conclude $|X \cap PY| = 0$.

Applying a character of order 8 to the equation

$$w_U w_U^{(-1)} = 64\left(1 + \frac{2^{2t-1}+1}{3} G/U\right) \tag{3.12}$$

following from Lemma 1.3.2, we see that $|X_0| \geq 1$. Furthermore,

$$\begin{aligned} |G/U| &= 8(2^{2t}-1)/3, \\ \sum a_g &= 8(2^{2t+1}+1)/3, \\ \sum a_g^2 &= 64\left(1 + \frac{2^{2t-1}+1}{3}\right). \end{aligned}$$

The formula for $\sum a_g^2$ follows by comparing the coefficient of 1 in (8).
We define $b_g := a_g - 2$. Then a calculation using the formulae for $|G/U|$, $\sum a_g$ and $\sum a_g^2$ gives

$$\sum b_g^2 = 64. \tag{3.13}$$

If $|X| \geq 2$ then $\sum b_g^2 \geq 72$ which is impossible. Thus $|X| = 1$. Let z be the element of order 2 of $\mathbb{Z}_8$. Since $|X \cap PY| = 0$ we conclude that

(*) for every complement U of $\mathbb{Z}_8$ in S_2 there is a coset L_U of $U\langle z\rangle$ such that one coset of U in L_U is completely contained in D and the other has empty intersection with D.

Write $H = \langle g_1, g_2\rangle \times K$ where possibly $|K| = 1$. Let $U_{ij} = \langle g_1 z^i, g_2 z^j\rangle \times K$, $i, j = 0, 1$. By (*), we have $L_{U_{ij}} \neq L_{U_{i'j'}}$ for $(i,j) \neq (i',j')$. Furthermore, the cosets $L_{U_{ij}}$, $(i,j) \neq (0,0)$, correspond to six coefficients 4 in w_H since every $L_{U_{ij}}$, $(i,j) \neq (0,0)$, is the union of two cosets of H which both intersect each of the two cosets of U_{ij} in $L_{U_{ij}}$ in exactly 4 elements.
Now, we will derive a contradiction to (3.13) for $U = H$. We know from above that w_H has one coefficient 8 and at least 6 coefficients 4. Let $\{a_g : g \in T\}$ be the remaining coefficients of w_H. Since

$$\begin{aligned} \sum_{g\in T} b_g &= |w_H| - (8 + 4\cdot 6) - 2(|G/H| - 7) \\ &= -10 \end{aligned}$$

we infer $\sum_{g\in T} b_g^2 \geq 10$. Thus

$$\begin{aligned} \sum b_g^2 &\geq (8-2)^2 + 6(4-2)^2 + 10 \\ &= 70, \end{aligned}$$

a contradiction to (3.13). □

3.2 Field descent exponent bounds

Now we are going to utilize the idea of the field descent from Chapter 2 to obtain a new type of exponent bound for difference sets and related objects which is more general than all previously known results.
In Section 2.3, the method of the field descent was used to derive a general bound on the absolute value of cyclotomic integers. This bound has strong implications on virtually all problems accessible to the character method. We will obtain strong asymptotic exponent bounds for groups containing difference sets without any restrictive assumptions like self-conjugacy. In many cases, previously literally nothing had been known on the existence of

these difference sets. Our results are a major step towards two longstanding open problems in difference sets, namely Ryser's conjecture and the circulant Hadamard matrix conjecture.
Furthermore, we will derive a general exponent bound on groups containing relative difference sets. As a consequence, we obtain strong necessary conditions for the existence of quasiregular projective planes which, in particular, lead to an asymptotic exponent bound for abelian groups admitting planar functions. Finally, we will also apply the field descent to group invariant weighing matrices.

3.2.1 A general exponent bound for difference sets

In this section, we derive a strong exponent bound for abelian groups containing difference sets. Our result does not rely on any restrictive assumption such as self-conjugacy and therefore is more general than all previously known nonexistence results on difference sets.
For a (v, k, λ, n)-difference set D in an abelian group G define

$$f(D) := \min\{t : \ \chi(D)\xi_v^{j(\chi)} \in \mathbb{Z}[\xi_t] \text{ for some } j(\chi) \text{ for all } \chi \in G^*\}.$$

In other words, $f(D)$ is the the smallest positive integer such that up to multiplication with a root of unity all character values of D lie in the $f(D)$th cyclotomic field. The results of this section will show that the parameter $f(D)$ is of fundamental importance for the study of difference sets. It is a striking fact that $f(D) = 1$ for all known difference sets with $\gcd(v, n) > 1$ in abelian groups. However, I am not aware of any previous general results on $f(D)$ besides the self-conjugacy condition which guarantees $f(D) = 1$, but does not apply in most cases.
We first state the most general version of our exponent bound. The main aim of this section being the study of difference sets in abelian groups, we also obtain a strong nonexistence result on difference sets in nonabelian groups as a by-product.

Theorem 3.2.1 *Assume the existence of a (v, k, λ, n)-difference set D in a group G. If U is a normal subgroup of G such that G/U is cyclic of order e then*

$$e \le \frac{vF(e,n)}{2\sqrt{n\varphi(F(e,n))}}$$

where φ denotes the Euler totient function and F is defined as in 2.2.5.

Proof Let $\rho : G \to G/U$ denote the canonical epimorphism. Then $E := \rho(D) = \sum_{i=0}^{e-1} a_i g^i$ where $0 \le a_i \le v/e$ and g is an element of G/U of order e. Let χ be the character of G/U defined by $\chi(g) = \xi_e$. Then $\chi(E) = \sum_{i=0}^{e-1} a_i \xi_e^i$ and $\chi(E)\overline{\chi(E)} = n$ by Lemma 1.3.7. Theorem 2.3.2 gives

$$n \le \frac{v^2 F(e,n)^2}{4e^2\varphi(F(e,n))}$$

implying the assertion. □

It is worth to state the abelian case separately.

Theorem 3.2.2 *Assume the existence of a (v,k,λ,n)-difference set in an abelian group G. Then*

$$\exp G \le \frac{vF(v,n)}{2\sqrt{n\varphi(F(v,n))}}.$$

In particular, if G is cyclic, then

$$n \le \frac{F(v,n)^2}{4\varphi(F(v,n))}.$$

Remark 3.2.3 In order to understand the strength of Theorem 3.2.2 we once more resort to an intuitive argument. Consider difference set parameters (v,k,λ,n) and write $v = \prod p_i^{a_i}$ where the p_i are prime. Define $f := \prod p_i$. For many parameters of putative difference sets and <u>all</u> parameter series of known difference sets with $\gcd(v,n) > 1$ we have $n \approx v$ and $f \approx \sqrt{v}$ or less. (again, "$\approx$" is not used in a strict sense here). For our reasoning we assume the worst case, i.e., $f \approx \sqrt{v}$. By Remark 2.2.9, we should have $F(v,n) \approx f$ in almost all cases. For a positive integer x, the order of magnitude of $x/\varphi(x)$ is $\pi^2/6$ on average, see [54]. So we consider $F(v,n)/\varphi(F(v,n))$ as a constant and get $F(v,n)/(2\sqrt{n\varphi(F(v,n))}) \approx \sqrt{F(v,n)/v}$. Using $n \approx v$ and $F(v,n) \approx \sqrt{v}$ we conclude that, loosely speaking, Theorem 3.2.2 shows

$$\exp G \le |G|^{3/4}$$

in almost all cases with $f \approx \sqrt{v}$ and $n \approx v$.

3.2.2 Difference sets with $\gcd(v,n) > 1$

The most interesting test cases for our exponent bound are the parameter series corresponding to known families of difference sets. In this section, we apply Theorem 3.2.1 to all parameter series corresponding to known difference sets with $\gcd(v,n) > 1$. The following is a complete list of these series, see [64, 65].

(i) Hadamard parameters:
$(v,k,\lambda,n) = (4u^2, 2u^2-u, u^2-u, u^2)$
where u is any positive integer.

(ii) McFarland parameters:
$(v,k,\lambda,n) = (q^{d+1}[\frac{q^{d+1}-1}{q-1}+1], q^d\frac{q^{d+1}-1}{q-1}, q^d\frac{q^d-1}{q-1}, q^{2d})$

where $q = p^f \neq 2$ and p is a prime.

(iii) Spence parameters:
$(v, k, \lambda, n) = (3^{d+1}\frac{3^{d+1}-1}{2}, 3^d\frac{3^{d+1}+1}{2}, 3^d\frac{3^d+1}{2}, 3^{2d})$
where d is any positive integer.

(iv) Chen/Davis/Jedwab parameters:
$(v, k, \lambda, n) = (4q^{2t}\frac{q^{2t}-1}{q^2-1}, q^{2t-1}[\frac{2(q^{2t}-1)}{q+1} + 1], q^{2t-1}(q-1)\frac{q^{2t-1}+1}{q+1}), q^{4t-2})$
where $q = p^f$, p is a prime, and t any positive integer.

We do not allow $q = 2$ for the McFarland parameters since then $(v, k, \lambda, n) = (2^{2d+2}, 2^{2d+1} - 2^d, 2^{2d} - 2^d, 2^{2d})$, and these are Hadamard parameters with $u = 2^d$. Hadamard difference sets are known to exist for every u of the form $u = 2^a 3^b r^2$ where $a, b \in \{0, 1\}$ and r is any positive integer, see [65]. Here we will consider arbitrary u. McFarland and Spence difference sets are known for any prime power q and any positive integer d, see [65]. Difference sets of type (iv) are known to exist only if f is even or $p \leq 3$, see [22, 33, 65]. However, here we will consider arbitrary f and p. We will first deal with Hadamard difference sets. A lot of work has been devoted to finding necessary conditions for the existence of Hadamard difference sets, see [2, 20, 21, 34, 96, 97, 98, 108, 127]. However, all these results rely either on the self-conjugacy condition or on very restrictive assumptions on the parameter u. In particular, almost nothing had been known on the existence of Hadamard difference sets when u has many prime divisors. The following consequence of our exponent bound 3.2.2 changes this situation dramatically.

Theorem 3.2.4 *For any finite set P of primes there is a computable constant $C(P)$ such that*

$$\exp G \leq C(P)|G|^{1/2}$$

for any abelian group G containing a Hadamard difference set whose order u^2 is a product of powers of primes in P.

Proof By Proposition 2.2.7 there is a constant $C_0(P)$ such that $F(4u^2, u^2) \leq C_0(P)$ for all u which are products of powers of primes in P. Also note that there is a constant E such that $x/\varphi(x) < E$ for all x whose prime divisors are all from P. Thus by Theorem 3.2.2, there is a constant $C(P)$ with $\exp G \leq C(P)u$. □

For a further asymptotic application of Theorem 3.2.2, we need the following well known result from analytic number theory, see [54, Thm. 328].

Result 3.2.5 *Let φ denote the Euler totient function. We have*

$$\limsup_{t \to \infty} \frac{t}{\varphi(t) \log \log t} = e^C$$

where $C \approx 0.5772$ is the Euler constant.

Theorem 3.2.6 *Let G be an abelian group containing a difference set whose parameters (v,k,λ,n) are of type* (ii), (iii) *or* (iv) *of the the above list. Then the following hold if $|G|$ is large enough.*

a)

$$\exp G < \left(\frac{\log\log 2p^{fd+1}}{p^{fd-1}}\right)^{1/2} |G|$$

for McFarland parameters,
b)

$$\exp G < \left(\frac{\log\log \frac{3^{d+2}-3}{2}}{3^{d-1}}\right)^{1/2} |G|$$

for Spence parameters,
c)

$$\exp G < \left(\frac{4\log\log 8p^{f(2t-1)+1}}{p^{f(2t-1)-1}}\right)^{1/2} |G|$$

for Chen/Davis/Jedwab parameters.

Proof a) Assume the existence of a difference set with McFarland parameters in an abelian group. We only deal with the case where p is odd. The case $p=2$ is similar. Taking $p_1 = p$ in Definition 2.2.5, we see that $f := F(v,n)$ divides $p(\frac{q^{d+1}-1}{q-1}+1)$ since $b_1 = 1$. This implies

$$f \leq 2p^{fd+1}. \tag{3.14}$$

Result 3.2.5 implies

$$\frac{f}{\varphi(f)} < 2\log\log 2p^{fd+1} \tag{3.15}$$

if $|G|$ and thus p^{fd} is large enough. Combining (3.14), (3.15) and Theorem 3.2.2 yields the assertion. The proofs for parts b) and c) are similar. □

3.2.3 Towards Ryser's conjecture

Ryser's conjecture [112, p. 139] asserts that there is no (v,k,λ,n)-difference set with $\gcd(v,n) > 1$ in any cyclic group. The next application of Theorem 3.2.2 shows that Ryser's conjecture is true for most of the parameters of known difference sets.

Theorem 3.2.7
a) If there is a Hadamard difference set in a cyclic group of order $v = 4u^2$ then $F(v,u)^2/\varphi(F(v,u)) \geq v$.

b) If there is a difference set with McFarland parameters in a cyclic group of order $q^{d+1}[\frac{q^{d+1}-1}{q-1}+1]$, $q = p^f$, *then* $p > 2$, $d = f = 1$ *and*

$$\frac{p+2}{\varphi(p+2)} \geq 4 - \frac{12}{p+2} \tag{3.16}$$

In particular, $p+2$ *has at least 20 distinct prime divisors and* $p > 2 \cdot 10^{28}$.
c) There are no difference sets with Spence or Chen/Davis/Jedwab parameters in any cyclic groups.

Proof a) This is immediate from Theorem 3.2.2.
b) Assume the existence of a difference set with McFarland parameters in a cyclic group G of order $v = q^{d+1}[\frac{q^{d+1}-1}{q-1}+1]$ where $q = p^f$, and p is a prime. We first show $f = d = 1$.
If we take $p_1 = p$ in Definition 2.2.5 then $b_1 = 1$ if p is odd and $b_1 = 2$ if $p = 2$. In both cases $f := F(v,n)$ divides $p(\frac{q^{d+1}-1}{q-1}+1)$ since $\frac{q^{d+1}-1}{q-1}+1$ is even for $p = 2$. Thus

$$f \leq 2p^{fd+1}. \tag{3.17}$$

Since $2 \cdot 3 \cdot 5/(1 \cdot 2 \cdot 4) < 4$, and since $r^{2/25} > r/(r-1)$ for all $r \geq 7$, we have

$$\frac{x}{4\varphi(x)} < x^{2/25}$$

for all integers $x > 1$. From Thereom 3.2.2 and (3.2.7) we thus get

$$p^{2fd} < (2p^{fd+1})^{27/25}.$$

This implies $fd = 1$ or $fd = 2$ and $p = 2$. In the latter case we have $f = 2$ and $d = 1$ since we assumed $q = p^f \neq 2$ for McFarland parameters. A direct application of Theorem 3.2.2 shows that this case cannot occur. Thus we have shown $fd = 1$.
Now let $fd = 1$. Then $p \neq 2$ since $q \neq 2$, and we have $v = p^2(p+2)$. Thus $f := F(v,n)$ divides $p(p+2)$. Theorem 3.2.2 gives $p^2 \leq p^2(p+2)^2/[4\varphi(p(p+2))]$ proving (3.16). Let $Y = 3 \cdot 5 \cdots 73$ be the product of the 20 smallest odd primes. Then $Y/\varphi(Y) < 3.97$ and $Y > 2 \cdot 10^{28}$. This implies the remaining assertions of part b.
c) In the case of Spence parameters we have $F(v,n) \leq 3(3^{d+1}-1)/2$ and $s \leq d$ in Theorem 3.2.2 and thus $\exp G < (2/3)^{d-2}v$. This leaves only the case $d = 1$ where we have $(v,k,\lambda,n)) = (36,15,6,9)$. But 3 is self-conjugate modulo 36, and thus no difference set with these parameters can exist in the cyclic group by Result 3.1.2.
In order to prove the nonexistence of difference sets with Chen/Davis/Jedwab parameters in cyclic groups G we apply Result 3.1.2. First assume that p is odd. Denote the Sylow p-subgroup of G by S_p. Note that S_p is cyclic of order

q^{2t}. Let U be the subgroup of G of order $2\frac{q^{2t}-1}{q^2-1}$. Then p is self-conjugate modulo $e := \exp G/U$ since e is 2 times a power of p. Thus Result 3.1.2 implies

$$q^{2t} = \exp S_p \leq |U||S_p|/q^{2t-1} = 2q\frac{q^{2t}-1}{q^2-1}.$$

Thus $1 < 2q/(q^2-1)$ contradicting $q \geq 3$.
Finally, for $p = 2$ we take $|U| = \frac{q^{2t}-1}{q^2-1}$, and apply the same argument. □

3.2.4 Circulant Hadamard matrices and Barker sequences

We recall that the **circulant Hadamard matrix conjecture** asserts that there is no circulant Hadamard matrix of order greater than 4. By Lemma 1.3.11, the order of a circulant Hadamard matrix is an even square, say $4u^2$. Futhermore, a circulant Hadamard matrix of order $4u^2$ exists if and only if there is a Hadamard difference set in the cyclic group of order $4u^2$. Turyn [127, 128] proved that u must be odd if a circulant Hadamard matrix of order $4u^2$ exists and that the circulant Hadamard matrix conjecture is true for $u < 55$. However, since Turyn's work in the 60s there has not been any progress on this conjecture because of the lack of methods to overcome the failure of the self-conjugacy approach.
Using the equivalence to Hadamard difference sets in cyclic groups, we can apply Theorem 3.2.7 a) to the circulant Hadamard matrix conjecture. Remark 2.2.8 strongly suggests that part a) of Theorem 3.2.7 should rule out the existence of circulant Hadamard matrices for almost all u. We confirm this by a computer search. Combining [127, Thm. 6] with Theorem 3.2.7 a, we find the following.

Theorem 3.2.8 *There is no circulant Hadamard matrix of order v in the range $4 < v \leq 10^{11}$ with the possible exceptions of $v = 4u^2$ with $u \in \{165, 11715, 82005\}$.*

A **Barker sequence** of length l is a sequence $(a_i)_{i=1}^{l}$ with $a_i = \pm 1$ such that $|\sum_{i=1}^{l-k} a_i a_{i+k}| \leq 1$ for $1 \leq k \leq l-1$. It is known that the existence of a Barker sequence of length $l > 13$ implies the existence of a circulant Hadamard matrix of order l, see [127, 128]. Thus $l = 4u^2$ where u is odd. It is shown in [42] that l cannot have a prime divisor $p \equiv 3 \pmod 4$ if $l > 13$ is the length of a Barker sequence. Combining these two results with Theorem 3.2.7 a) we get the following bound by a computer search. It improves the previously known bound [41, p. 363] by a factor greater than 10^6. We do <u>not</u> need Turyn's bound [127, Thm. 6] to obtain this result.

Theorem 3.2.9 *There is no Barker sequence of length l with*

$$13 < l \leq 4 \cdot 10^{12}.$$

3.2.5 Relative difference sets and planar functions

In this section, we utilize the method of the field descent to obtain a nonexistence theorem on relative difference sets. No results of comparable generality had previously been known. A treatment of most of the previously known results can be found in [105, Chapters 4,5]. In particular, we will obtain new necessary conditions for the existence of relative $(n, n, n, 1)$-difference sets which are equivalent to quasiregular projective planes of type b) of the Dembowski/Piper classification [37]. We will combine this result with a further new nonexistence theorem on relative $(n, n, n, 1)$-difference sets to derive a strong asymptotic exponent bound on abelian groups admitting planar functions.

If the prime power conjecture for projective planes is true, then, in particular, n must be a prime power if a relative $(n, n, n, 1)$-difference set exists. It is known that n must be a power of 2 if a relative $(n, n, n, 1)$-difference set with even n exists in an abelian group, see [45] or [63]. In an important paper, Ma [83] proved that there is no relative $(n, n, n, 1)$-difference set in an abelian group if n is a product of two primes. However, aside from Ma's result and a simple exponent bound [105, Thm. 4.1.1] very little had been known about the existence of relative $(n, n, n, 1)$-difference sets in general – especially if n has many prime divisors, see [105, Section 5.4]. As for the Hadamard difference sets, our results are the first to tackle these cases.

Theorem 3.2.10 *Assume the existence of an (m, n, k, λ)-difference R set in a group G relative to N. Let U be any subgroup of G not containing N such that G/U is cyclic of order e. Then*

$$|U \cap N| \leq \frac{|U|F(e,k)}{2\sqrt{k\varphi(F(e,k))}}.$$

Proof Let $\rho : G \to G/U$ be the canonical epimorphism, and let χ be a character of G/U of order e. Note that χ is nontrivial on $N_U := NU/U$ since U does not contain N. Since any coset of N contains at most one element of R and since χ has a trivial kernel, we have $\chi(\rho(R)) = \sum_{i=0}^{e-1} a_i \xi_e^i$ with $0 \leq a_i \leq C$ where $C = |U|/|U \cap N|$. Since χ is nontrivial on N_U we get

$$k = \chi(\rho(R))\overline{\chi(\rho(R))} \leq \frac{|U|^2 F(e,k)^2}{4\varphi(F(e,k))|U \cap N|^2}$$

from Lemma 1.3.6 and Theorem 2.3.2 proving the theorem. □

Now we are going to study relative $(n, n, n, 1)$-difference sets corresponding to quasiregular projective planes of type b) of the Dembowski/Piper classification [37], see Proposition 3.2.11 below. These projective planes (of order n) admit a quasiregular collineation group of order n^2 with exactly three point orbits whose sizes are $1, n, n^2$. Here a collineation group G is called **quasiregular** if it induces a regular operation on all its point orbits, i.e., if all points in any fixed orbit of G have the same stabilizer. Since the conjugates of a point stabilizer coincide with the stabilizers of the points in the same orbit, a collineation group G is quasiregular if and only if all its point stabilizers are normal subgroups of G. In particular, any abelian collineation group is quasiregular. Next, we describe the connection between quasiregular projective planes and relative $(n, n, n, 1)$-difference sets. For the convenience of the reader, we sketch the proof.

Proposition 3.2.11 *There is a projective plane of order n with a quasiregular collineation group G of order n^2 and point orbits of size $1, n, n^2$ if and only if there is an $(n, n, n, 1)$-difference set R in G relative to a normal subgroup N.*

Proof Assume that there is a projective plane of order n with a collineation group G as described in the assertion. By [37, Thm. 4] the orbits of size $1, n$ form an incident point-line pair (p_0, L_0). Since G acts regularly on the point orbit $\mathcal{O}$ of size n^2, we may identify G with $\mathcal{O}$. Let $p \neq p_0$ be a point incident with L_0, and let $L \neq L_0$ be a line through p. Then $N := G_p$ is a normal subgroup of G of order n, and a straightforward verification shows that $L \setminus \{p\}$ is an $(n, n, n, 1)$-difference set in G relative to N. The converse is proven by reversing this construction. □

The next theorem will be needed for the proof of our asymptotic exponent bound for groups admitting planar functions.

Theorem 3.2.12 *Let G be an abelian group containing an $(n, n, n, 1)$-difference set R relative to N. Let p be a prime divisor of n, and let S be the Sylow p-subgroup of N. If p^a is the exact power of p dividing n then*

$$\exp S \leq p^{\lceil a/2 \rceil}$$

where $\lceil x \rceil$ denotes the smallest integer $\geq x$.

Proof
Let $o(g)$ denote the order of an element g of G. Assume $e := \exp S \geq p^{\lceil a/2 \rceil + 1}$ and let $S = \langle a_1 \rangle \times \cdots \times \langle a_t \rangle$, $o(a_1) = e$, be a decomposition of S into a direct product of cyclic groups. Let $G = \langle b_1 \rangle \times \cdots \times \langle b_s \rangle$ be a decomposition of G into cyclic groups of prime power order and write $a_1 = \prod_{j=1}^{s} b_j^{t_j}$ where w.l.o.g. $o(b_1^{t_1}) = e$.
Define $\chi \in G^*$ by $\chi(b_1) = \xi_{o(b_1)}$ and $\chi(b_j) = 1$ for $j > 1$ and write $K :=$ Ker χ. Then G/K is a cyclic p-group whose order is at least $p^{\lceil a/2 \rceil + 1}$ since $o(b_1) \geq o(b_1^{t_1}) = e \geq p^{\lceil a/2 \rceil + 1}$. Furthermore, $|S \cap K| = |\mathrm{Ker}\ \chi|_S| = p^a/e$.
Let $\rho : G \to G/K$ be the canonical automorphism. We have $\chi(\rho(R))\overline{\chi(\rho(R))} \in \{0, n\}$ by Lemma 1.3.6 implying $\chi(\rho(R))\overline{\chi(\rho(R))} \equiv 0 \pmod{p^a}$ for every nontrivial $\chi \in (G/K)^*$. Since p is self-conjugate modulo any power of p and thus modulo $\exp G/K$, we get $\chi(\rho(R)) \equiv 0 \pmod{p^{\lfloor a/2 \rfloor}}$ for every nontrivial

$\chi \in (G/K)^*$ from Corollary 1.4.5. This congruence also holds for the trivial character χ_0 of G/K since $\chi_0(\rho(R)) = |R| = n$. Thus we can apply Lemma 1.5.2 and get (using the notation of Lemma 1.5.2)

$$\rho(R) = p^{\lfloor a/2 \rfloor} X_0 + p^{\lfloor a/2 \rfloor - 1} P_1 X_1 + \cdots + X_{\lfloor a/2 \rfloor} P_{\lfloor a/2 \rfloor}.$$

Thus $p^{\lfloor a/2 \rfloor}$ divides $\rho(R)\rho(R)^{(-1)}$. From Lemma 1.3.1 we get

$$\rho(R)\rho(R)^{(-1)} = n - |S \cap K|N + |K|G.$$

Thus $p^{\lfloor a/2 \rfloor}$ divides $|S \cap K| = p^a/e$ contradicting $e \geq p^{\lceil a/2 \rceil + 1}$. □

It is known that a planar function from $\mathbb{Z}_n$ to $\mathbb{Z}_n$ cannot exist if n is even, not squarefree or the product of two primes or if there are two prime divisors p, q of n such that p is self-conjugate modulo q, see [83, Thm. 1.1, Cor. 4.4]. However, very little has been known about planar functions $f : H \to N$ for which H and N are noncyclic abelian groups. Our next result provides an asymptotic exponent bound on H and N.

Theorem 3.2.13 *For any finite set P of primes there is a computable constant $C(P)$ such that*

$$\exp H, \exp N \leq C(P)\sqrt{n}$$

for any abelian groups H, N admitting a planar function $f : H \to N$ whose degree n is a product of powers of primes in P.

Proof Assume that there is a planar function $f : H \to K$ where H and K are abelian groups of order n, and n is product of powers of primes in P. By Lemma 1.2.8 there is an $(n, n, n, 1)$-difference set in $G := H \times N$ relative to N. From Theorem 3.2.12 we get

$$\exp N \leq \sqrt{n} \prod_{p \in P} \sqrt{p}. \tag{3.18}$$

Let χ be a character of G of order $e := \exp H$ with $|\ker \chi \cap N| = n/p$ where p is some prime divisor of n. Write $U := \ker \chi$ and note $|U| = n^2/e$. Then G/U is cylic of order e and thus

$$|U \cap N| = n/p \leq \frac{F(e,n)n^2}{2e\sqrt{n\varphi(F(e,n))}}$$

by Theorem 3.2.10. Hence, by Proposition 2.2.7,

$$e = \exp H \leq C'\sqrt{n} \tag{3.19}$$

for some constant C' only depending on P. Now the assertion follows from (3.18) and (3.19). □

3.2.6 Group invariant weighing matrices

Our last application of the field descent will be to group invariant weighing matrices. For the basics on group invariant weighing matrices, see Section 1.2.13. Very little has been known on the existence of group invariant weighing matrices. The case which has attracted the most attention is that of *circulant* weighing matrices, i.e., matrices $W(m,n)$ which are invariant under the cyclic group $\mathbb{Z}_m$, see [40, 101, 102, 120]. The nicest result in this direction is the following consequence of the solution of the so-called **Waterloo problem** due to Arasu, Dillon, Jungnickel, and Pott [4]. See also [105, Chapter 3.3].

Result 3.2.14 *Let q be a prime power, and let $d \geq 3$ be an integer. A circulant weighing matrix $W((q^d-1)/(q-1), q^{d-1})$ exists if and only if d is odd.*

Note that if d is even, say $d = 2a$, there is a subgroup of order $q^a + 1$ of $\mathbb{Z}_{(q^d-1)/(q-1)}$, and p is self-conjugate modulo $q^a + 1$. Using [127, Thm. 6], it is easy to prove the nonexistence part of Result 3.2.14. The existence part was established in [4] by a construction using quadrics in projective geometries. There are some further nonexistence results for circulant weighing matrices. It has been shown that there are no circulant weighing matrices $W(m, m-1)$ for $m > 2$ [102] and that a circulant weighing matrix $W(m,n)$ with odd m can only exist if $(m-n)^2 - (m-n) \geq n-1$ [46]. Further nonexistence results can be obtained using multiplier theorems or Turyn's self-conjugacy approach. However, these methods only work under severe restrictions on the parameters m and n and, as usual, fail in most cases when m or n have many prime divisors. As a consequence of Theorem 2.3.2 we obtain the following result which is of much broader applicability. Recall that by Lemma 1.2.13 a group invariant weighing matrix $W(m,n)$ can only exist if n is a square.

Theorem 3.2.15 *Assume the existence of a G-invariant weighing matrix $H = W(m,s^2)$ where s is a positive integer. Let U be a subgroup of G such that G/U is cyclic of order e. Then*

$$s \leq \frac{|U|F(m,s)}{\sqrt{\varphi(F(m,s))}}.$$

In particular, the existence of a circulant weighing matrix $W(m,s^2)$ implies

$$s \leq \frac{F(m,s)}{\sqrt{\varphi(F(m,s))}}.$$

Proof Let $\rho : G \to G/U$ be the canonical epimorphism, and let χ be a character of G/U of order e. If we view H as an element of $\mathbb{Z}[G]$ (see Lemma 1.3.9) then $\rho(H) = \sum_{g \in G/U} a_g g$ with $|a_g| \leq |U|$ for all g since H

has coefficients $-1,0,1$ only. As χ has a trivial kernel, we get $\chi(\rho(H)) = \sum_{i=0}^{e-1} b_i\xi_e^i$ with $|b_i| \le |U|$ for all i. Adding $0 = |U|\sum_{i=0}^{e-1}\xi_e^i$, we get $\chi(\rho(H)) = \sum_{i=0}^{e-1} a_i\xi_e^i$ with $0 \le a_i \le 2|U|$. Now we apply Lemma 1.3.10 and Theorem 2.3.2 and get

$$s^2 = \chi(\rho(H))\overline{\chi(\rho(H)))} \le |U|^2 F(e,s)^2/\varphi(F(e,s))$$

(note $F(e,s) = F(e,s^2)$) proving the assertion. □

Note that Theorem 3.2.15 is weaker than Theorem 3.2.7 a) in the case of *circulant* Hadamard matrices since we had to deal with coefficients $-1,0,1$ instead of just $-1,1$. As an example illustrating the power of Theorem 3.2.15 we give an application to the family of group invariant weighing matrices $W(2s^2,s^2)$ where s is a positive integer. This is a rich and interesting family since examples for such matrices are known for any square s: There are Hadamard difference sets of order $n = s^2$ in suitable abelian groups G for any square s [22]. If D is such a Hadamard difference set (viewed as a group ring element) and $\rho : G \to G/U$ is a projection onto a subgroup U of G of order 2 then $\rho(D) - (G/U) \in \mathbb{Z}[G/U]$ corresponds to a G/U-invariant weighing matrix $W(2s^2,s^2)$. It is straightforward to verify this using Lemmas 1.3.2 and 1.3.9.

Corollary 3.2.16 *Let P be any finite set of primes, and let Q be the set of all products of powers of primes in P. Then there is a computable constant $C(P)$ such that*

$$\exp(G) \le C(P)s$$

for any $s \in Q$ and any abelian group G of order $2s^2$ for which a G-invariant weighing matrix $W(2s^2,s^2)$ exists.
In particular, a circulant weighing matrix $W(2s^2,s^2)$ can only exist for finitely many $s \in Q$.

Proof This is immediate from Proposition 2.2.7 and Theorem 3.2.15. □

Chapter 4

Two-weight irreducible cyclic codes

The determination of the weight distribution of irreducible cyclic codes is a fascinating problem which can be tackled by an interplay of number theoretic and combinatorial techniques. Important contributions in this direction can be found in [9, 11, 71, 93]. An identity due to McEliece [93] shows that the weights of an irreducible cyclic code can be expressed by linear combinations of Gauss sums via the Fourier transform. This makes number theoretic techniques available for the determination of the weights of irreducible cyclic codes. However, McEliece's identity also indicates that this problem is extremely difficult in general since the same is true for the evaluation of Gauss sums.

Even the *two-weight* irreducible cyclic codes have not yet been classified. Two infinite families of two-weight irreducible cyclic codes and seven sporadic examples are known, see Section 4.2. In this chapter, we will give a unified explanation for all these two-weight codes, find four new sporadic examples and provide evidence that there are no further examples.

The main point of our approach is to find "simple" necessary and sufficient numerical conditions for an irreducible cyclic code to have at most two weights. In Section 4.1, we will derive these conditions without evaluating the corresponding Gauss sums in McEliece's identity; we only need to use the factorization of Gauss sums given by Stickelberger's theorem and Parseval's identity for Fourier transforms. What makes the analysis of our "simple" conditions complicated is that they involve a parameter θ coming from Stickelberger's theorem which behaves as irregular as class numbers of imaginary quadratic number fields. Nevertheless, we believe to have found *all* two-weight irreducible cyclic codes. The classification is described in detail in Section 4.2.

In Section 4.3, we prove the completeness of our classification in some cases by

resorting to an class number estimate conditionally on GRH due to Louboutin [81]. We also use recent results on Gauss sums due to Mbodj [92].
Two-weight irreducible cyclic codes can also be studied in terms of two-intersection sets in finite projective spaces and in terms of difference sets. Since these viewpoints are enlightening sometimes, we explain them in Section 4.4. The two-intersection sets corresponding to the eleven sporadic two-weight codes all have the interesting property that the square of the difference of their intersection numbers is not the order of the underlying geometry. Such sets are rare and have received some recent interest.
Turning to the difference set interpretation, we arrive at the most elegant way to phrase our results: subject to our conjecture, there are exactly eleven sub-difference sets of Singer difference sets which are not trivial or Singer difference sets themselves. We will identify these eleven examples among the known difference sets.

4.1 A necessary and sufficient condition

We now state and prove the necessary and sufficient numerical conditions on the parameters of an irreducible cyclic code to have at most two nonzero weights. As explained in Section 1.2.6, for the classification of two-weight irreducible cyclic codes, it suffices to consider the codes $C(q,m,u)$ defined below.

Definition 4.1.1 *Let L/K be an extension of finite fields of degree m where K has order q. Let n be a divisor of q^m-1 divisible by $q-1$, and write $u=(q^m-1)/n$. Let ω be a primitive nth root of unity in L. Define*

$$C(q,m,u) := \left\{ c(y) := \left(\mathrm{Tr}_{L/K}(y\omega^i)\right)_{i=0}^{n-1} \;\middle|\; y \in L \right\}.$$

By Proposition 1.2.16, the codes $C(q,m,u)$ are irreducible cyclic codes. We use the parameter u instead of n in the notation of these codes as u will play a more important role. We now show that we can restrict our attention to the codes $C(q,m,u)$ with q prime.

Proposition 4.1.2 *Let $q=p^t$ where p is a prime. Then $C(q,m,u)$ is a two-weight code if and only if $C(p,mt,u)$ is a two-weight code.*

Proof Let $K=\mathbb{F}_q$. Because of the K-linearity of $\mathrm{Tr}_{L/K}$ and since $K^* \subset \langle\omega\rangle$, the nonzero entries of the codeword $c(y) = (\mathrm{Tr}_{L/K}(y\omega^i))_{i=0}^{n-1}$ of $C(q,m,u)$ are uniformly distributed over K^*. Let $w(y)$ be the weight of $c(y)$, and let $w'(y)$ be the weight of the the corresponding codeword $(\mathrm{Tr}_{L/\mathbb{F}_p)}(y\omega^i))_{i=0}^{n-1}$ of $C(q,mt,u)$. Then $w'(y) = w(y) - [w(y)/(q-1)](q/p-1) = w(y)q(p-1)/[(q-1)p]$. Thus the weights of $C(q,m,u)$ and $C(p,mt,u)$ only differ by a constant factor. □

In view of Propostion 4.1.2, we will only consider codes $C(p,m,u)$ with p prime henceforth. As a preparation for the proof of our main theorem, we need a lemma on maximum p-powers dividing Gauss sums. We recall that for a character χ of $\mathbb{F}_q^*$, $q = p^f$, p prime, the corresponding Gauss sum is defined by

$$G(\chi) = \sum_{x \in \mathbb{F}_q^*} \chi(x) \xi^{\mathrm{Tr}_{\mathbb{F}_q/\mathbb{F}_p}(x)}.$$

As in Stickelberger's theorem, $S_p(y)$ will denote the sum of the p-digits of y.

Lemma 4.1.3 *Let p be a prime, and let u be a positive integer with $(u,p) = 1$. Write $f := \mathrm{ord}_u(p)$. Define*

$$\theta(u,p) := \frac{1}{p-1} \min\{S_p(\tfrac{j(p^f-1)}{u}) : 1 \le j < u\}.$$

Let s be a positive integer. If u divides $(p^{sf}-1)/(p-1)$, then $p^{s\theta(u,p)}$ is the largest p-power dividing $G(\chi)$ for every nontrivial character χ of $\mathbb{F}_{p^{sf}}^*$ such that χ^u is trivial.

Proof First of all, we can assume $s = 1$, since the general assertion follows from the case $s = 1$ and the Davenport-Hasse theorem 1.4.10. Write $q := p^f$. Let π be an arbitrary prime ideal above p in $\mathbb{Q}(\xi_{q-1})$ and let $\tilde{\pi}$ be a prime ideal above π in $\mathbb{Q}(\xi_{q-1}, \xi_p)$. By Lemma 1.4.2, we have $\pi = \tilde{\pi}^{p-1}$. Let τ be the Teichmüller character of $\mathbb{F}_q^*$ corresponding to π. Since τ has order $q-1$, the set of nontrivial characters of $\mathbb{F}_q^*$ of order dividing u is

$$\{\tau^{j(q-1)/u} : 1 \le j < u\} =: U.$$

Let $\nu_{\tilde{\pi}}$ denote the $\tilde{\pi}$-adic evaluation. By Stickerberger's theorem 1.4.9, we have

$$\nu_{\tilde{\pi}}(G(\tau^l)) = S_p(l) \tag{4.1}$$

for $l = 1, ..., q-2$. By Lemma 1.4.8, we have $G(\chi) \in \mathbb{Z}[\xi_{q-1}]$ for all $\chi \in U$. Thus, since $\pi = \tilde{\pi}^{p-1}$ and because of (4.1), we get $S_p(l) \equiv 0 \pmod{p-1}$ for $l = j(q-1)/u$, $1 \le j < u$. This shows that $\theta(u,p)$ is an integer. By (4.1), the maximum power of π dividing $G(\chi)$ for all $\chi \in U$ is

$$\frac{1}{p-1} \min\{\nu_{\tilde{\pi}}(G(\chi)) : \chi \in U\} = \theta(u,p).$$

Note that $\theta(u,p)$ does not depend on the prime ideal π. Thus, for any prime ideal π above p in $\mathbb{Q}(\xi_{q-1})$, the maximum power of π dividing $G(\chi)$ for all $\chi \in U$ is $\theta(u,p)$. Since p is unramified in $\mathbb{Q}(\xi_{q-1})/\mathbb{Q}$, this implies the assertion. □

Theorem 4.1.4 *Let p be a prime, and let u, m be positive integers such that u divides $(p^m-1)/(p-1)$. Write $\theta=\theta(u,p)$ and $m=fs$ with $f:=\mathrm{ord}_u(p)$. Then $C(p,m,u)$ is a two-weight code if and only if there exists a positive integer k satisfying*

$$\left.\begin{aligned} &k \mid u-1 \\ &kp^{s\theta} \equiv \pm 1 \pmod{u} \\ &k(u-k)=(u-1)p^{s(f-2\theta)} \end{aligned}\right\} \tag{4.2}$$

Proof *Necessity.* Assume that $C(p,m,u)$ is a two-weight code. Let ω be a primitive nth root of unity in $\mathbb{F}_{p^m}$ where $n=(p^m-1)/u$. By Proposition 1.2.19, the Singer difference set D of $\mathrm{PG}(m-1,p)$ has a sub-difference set E in $H:=G/\langle\omega\rangle$, where $G=\mathbb{F}_{p^m}^*/\mathbb{F}_p^*$ is the Singer cycle of $\mathrm{PG}(m-1,p)$. Since $|G/\langle\omega\rangle|=u$, the parameters of E are of the form (u,k,λ). By replacing E by its complement in H if necessary, we may assume $k<u/2$ and thus $\lambda<u/4$. We will show that k satisfies (4.2). Write $l:=k-\lambda$. Let $\rho:G\to H$ be the canonical epimorphism. Throughout this proof, characters χ of H are also viewed as characters of G via $\chi(g)=\chi(g\langle\omega\rangle)$ for $g\in G$. We have

$$\rho(D)=aE+bH \tag{4.3}$$

for some integers a,b by the definition of a sub-difference set. Let χ be a nontrivial character of H. Then $|\chi(D)|^2=p^{m-2}$ and $|\chi(E)|^2=l$ by Lemma 1.3.7. On the other hand, from (4.3) we get $\chi(D)=a\chi(E)$ and thus

$$p^{m-2}=a^2l. \tag{4.4}$$

Hence a and l is are both powers of p. This together with $\lambda<n$, $(u,p)=1$ and $k^2=l+\lambda u$ implies $(k,\lambda)=1$. Since $k(k-1)=\lambda(u-1)$, we see that k divides $u-1$.
By Lemma 1.4.8, we have $p\chi(D)=G(\chi)$ for all nontrivial characters χ of H. Thus, by Lemma 4.1.3, the maximum power of p dividing $\chi(D)$ for all nontrivial characters χ of H is $p^{s\theta-1}$. Since a is a power of p and $\chi(D)=a\chi(E)$ for all nontrivial characters χ of H, we conclude that a divides $p^{s\theta-1}$. We claim $a=\pm p^{s\theta-1}$. Otherwise a divides $p^{s\theta-2}$, and p divides $\chi(E)$ for all nontrivial characters χ of H since $\chi(D)=a\chi(E)$. By Corollary 1.3.5, this implies $E=pX+cH$ for some $X\in\mathbb{Z}[H]$ and some integer c. Note that $X\neq 0$ since otherwise $\chi(E)=0$ for all nontrivial characters χ of H. Since E has coefficients $0,1$ only, $E=pX+cH$ implies that X is a multiple of H. But then again $\chi(E)=0$ for all nontrivial characters χ of H, a contradiction. This shows $a=\pm p^{s\theta-1}$. Thus (4.3) now reads

$$\rho(D)=\pm p^{s\theta-1}E+bH. \tag{4.5}$$

Recall that $|\rho(D)| = |D| = (p^{m-1}-1)/(p-1)$. From (4.5) we get

$$p^{m-1} \equiv \frac{p^{m-1}-1}{p-1} \equiv |\rho(D)| \equiv \pm p^{s\theta-1}k \pmod{u}.$$

Hence $p^{s\theta}k \equiv \pm p^m \equiv \pm 1 \pmod{u}$ verifying the second condition in (4.2). From (4.4) and $a = \pm p^{s\theta-1}$ we get $l = p^{m-2s\theta}$. Hence $\lambda = k - p^{m-2s\theta}$. Thus $k(k-1) = \lambda(u-1)$, i.e., the trivial necessary condition for the existence of a (u,k,λ) difference set, implies the third condition in (4.2).

Sufficiency. Assume that (4.2) holds. Define D, H and ρ as in the first part of the proof. Define an element E of the group algebra $\mathbb{Q}[H]$ by

$$E := (\rho(D) - xH)/p^{s\theta-1} \tag{4.6}$$

where

$$x = \frac{(p^{m-1}-1) \mp k(p-1)p^{s\theta-1}}{u(p-1)}.$$

First of all, x is an integer because of the second condition in (4.2). Let χ_0 be the principal character of H.
Then $\chi_0(\rho(D) - xH) = (p^{m-1}-1)/(p-1) - [(p^{m-1}-1)/(p-1) \mp kp^{s\theta-1}] = \pm kp^{s\theta-1}$. Furthermore, all nontrivial character values of $\rho(D) - xH$ are divisible by $p^{s\theta-1}$ by Lemmas 1.3.7, 1.4.8 and 4.1.3. Thus $\rho(D) - xH$ is divisible by $p^{s\theta-1}$ by Lemma 1.3.4. Hence $E \in \mathbb{Z}[H]$. Write $E = \sum_{h\in H} e_h h$ with $e_h \in \mathbb{Z}$. Then $\sum e_h = \chi_0(E) = \pm k$. Furthermore, by Lemma 1.3.2 applied to D, we have

$$EE^{(-1)} = p^{m-2s\theta} + \frac{k^2 - p^{m-2s\theta}}{u} H.$$

Thus $\sum e_h^2 = (k^2 - (u-1)p^{m-2s\theta})/u = k$ using the third condition in (4.2). This implies $e_h = \pm 1$ for all $h \in H$. Thus E is a sub-difference set of D in H. Now the assertion follows from Proposition 1.2.19. □

4.2 All two-weight irreducible cyclic codes?

Using Theorem 4.1.4 we can attempt to classify all two-weight irreducible cyclic codes by finding all solutions to (4.2).

4.2.1 Subfield and semiprimitive codes

There are two known infinite families of all two-weight irreducible cyclic codes: the subfield codes and the semiprimitive codes. We now describe the corresponding solutions of (4.2). We use the notation of Definition 4.1.1. The most obvious two-weight codes $C(p,m,u)$ arise if ω generates a subfield of L.

Proposition 4.2.1 *If ω is a primitive element for a subfield $F \cong \mathbb{F}_{p^a}$ of $L = \mathbb{F}_{p^m}$, then $C(p,m,u)$ has only one nonzero weight.*

Proof Let $y \in L^*$. If $\mathrm{Tr}_{L/F}(y) = 0$, then $\mathrm{Tr}_{L/K}(y\omega^i) = 0$ for all i. If $\mathrm{Tr}_{L/F}(y) \neq 0$, then $\{\omega^i : \mathrm{Tr}_{L/K}(y\omega^i) = 0\} \cup \{0\}$ is a K-vector space of dimension $a-1$. Thus the only nonzero weight of $C(p,m,u)$ is $p^a - p^{a-1}$. □

We call the codes appearing in Proposition 4.2.1 subfield codes. From the proofs of Theorem 4.1.4 and Proposition 4.2.1 we see that $k = (p^{m-a} - 1)/(p^a - 1)$ in (4.2) and thus $\theta(u,p) = a$ for a subfield code $C(p,m,u)$. Thus we have the following.

Proposition 4.2.2 *The subfield codes $C(p,m,u)$ exactly correspond to the solutions of* (4.2) *with*

$$\begin{aligned} u &= (p^m - 1)/(p^a - 1) \\ k &= (p^{m-a} - 1)/(p^a - 1) \\ \theta(u,p) &= a. \end{aligned}$$

Now we come to the semiprimitive codes. A prime p is called **semiprimitive** modulo u if -1 is power of p modulo u. Note that (4.2) has a solution with $k =\in \{1, u-1\}$ if and only if $\theta(u,p) = f/2$. By [10, Thms. 1,4], we have $\theta(u,p) = f/2$ if and only if p is semiprimitive modulo u. Thus we have the following.

Proposition 4.2.3 *There is a solution of* (4.2) *with $k \in \{1, u-1\}$ if and only if p is semiprimitive modulo u. The corresponding two-weight codes $C(p,m,u)$ are called* **semiprimitive codes**.

4.2.2 The exceptional codes

Two-weight irreducible cyclic codes which are neither subfield nor semiprimitive codes will be called **exceptional**. The corresponding solutions of (4.2) will also be called **exceptional**. Theorem 4.1.4 makes possible a computer search for exceptional codes. This can be done as follows. For every proper divisor $k > 1$ of $u - 1$ compute $k(u-k)/(u-1)$. If it is a prime power, say p^r, check whether $f - 2\theta$ divides r. If so and the quotient is s, as long as the congruence condition of (4.2) holds, $C(p, fs, u)$ is a two-weight irreducible cyclic code. The following table lists all exceptional solutions of (4.2) with $u \leq 100,000$.

u	p	s	f	θ	k	ε
11	3	1	5	2	5	+1
19	5	1	9	4	9	+1
35	3	1	12	5	17	+1
37	7	1	9	4	9	+1
43	11	1	7	3	21	+1
67	17	1	33	16	33	+1
107	3	1	53	25	53	+1
133	5	1	18	8	33	−1
163	41	1	81	40	81	+1
323	3	1	144	70	161	+1
499	5	1	249	123	249	+1

The two-weight codes from above with $u \in \{11, 19, 67, 107, 163, 499\}$ were already found by Langevin [71]. His proof relies on the fact that the Gauss sums in McEliece's indentity can be evaluated if u is prime and $f = (u-1)/2$. Batten and Dover [7] verified by computer that $C(7, 9, 37)$ is a two-weight code. The result in [7] is presented as a certain two-intersection set in $\mathrm{PG}(2, 7^3)$; see Section 4.4 for more on the correspondence between two-weight codes and two-intersection sets. We believe that $C(3, 12, 35)$, $C(11, 7, 43)$, $C(5, 18, 133)$ and $C(3, 144, 323)$ are new two-weight codes.
The fact that there are no exceptional solutions with $500 \le u \le 100,000$ and the results of the next section provide evidence for the following.

Conjecture 4.2.4 *An irreducible cyclic code $C(p, m, u)$ is a two-weight code if and only if it is a subfield code, a semiprimitive code or appears in the table of exceptional codes above.*

4.3 Partial proof of Conjecture 4.2.4

Conditionally on GRH, we give a partial proof of Conjecture 4.2.4. Again we only consider codes $C(p, m, u)$ with p prime.
One of the tools we will need is a bound on class numbers of imaginary quadratic fields due to Louboutin [81]. Let K be an imaginary quadratic number field, and let $\zeta_K(s)$ denote its Dedekind zeta function, see [13, p. 309]. We recall that the generalized Riemann hypothesis (GRH) for K asserts that $\Re s = 1/2$ for all zeros s of $\zeta_K(s)$ with $0 < \Re s < 1$. Here $\Re s$ denotes the real part of the complex number s.

Result 4.3.1 (Louboutin [81]) *Let u be a square-free positive integer and let $h(-u)$ denote the class number of $K = \mathbb{Q}(\sqrt{-u})$. Assuming GRH for K, we have*

$$h(-u) \ge \frac{\pi\sqrt{u}}{3e \log u}.$$

We combine Louboutin's bound with recent work of Langevin [72] and Mbodj [92] on Gauss sums to prove the following.

Theorem 4.3.2 *Conditionally on GRH, there are no two-weight irreducible cyclic codes $C(p,m,u)$ for which the triple (p,m,u) satisfies any of the following conditions.*

(a) $u \equiv 0 \pmod 3$, $u \neq 3$, $p \equiv 1 \pmod 3$ *and*

$$m > \frac{3\log((u+1)/4)}{\log p}. \tag{4.7}$$

(b) There is a prime divisor $r \equiv 3 \pmod 4$ *of* u *with* $r > 3$,

$$\mathrm{ord}_r(p) = (r-1)/2 \tag{4.8}$$

and

$$m > \frac{3e(r-1)\log r \log((u+1)/4)}{2\pi\sqrt{r}\log p}. \tag{4.9}$$

(c) There are two odd prime divisors $r,\ s > 3$ *of* u *such that*

$$\mathrm{ord}_r(p) = r-1, \mathrm{ord}_{rs}(p) = (r-1)(s-1)/2 \tag{4.10}$$

and

$$m > \frac{3e(r-1)(s-1)\log rs \log((u+1)/4)}{2\pi\sqrt{rs}\log p}. \tag{4.11}$$

Proof (b) Assume that $C(p,m,u)$ is a two-weight code. Write $f = \mathrm{ord}_u(p)$, $m = ft$, $g = (r-1)/2$, and let χ be a character of $\mathrm{GF}(p^g)$ of order r. By [72], the exact power of p dividing the Gauss sum $G(\chi)$ is $p^{(g-h)/2}$ where h is the class number of $\mathbb{Q}(\sqrt{-r})$. Thus, by the Davenport-Hasse theorem and Lemma 4.1.3, $2\theta(u,p) \leq f - hf/g$. Recall that $k(u-k) = (u-1)p^{t(f-2\theta(u,p))}$ for some divisor k of $u-1$ by Theorem 4.1.4. Note that $k(u-k)/(u-1) \leq (u+1)/4$. Putting this together, we get

$$\frac{u+1}{4} \geq p^{t(f-2\theta(u,p))} \geq p^{mh/g}.$$

Now assertion (b) follows by taking logarithms and using Result 4.3.1.
The proof of part (c) is similar. If $s \equiv 3 \pmod 4$ and $\mathrm{ord}_s(p) = (s-1)/2$ then the bound from part (b), with s in place of r, implies the bound in (c). Otherwise, we may use Proposition 3.8 of [92] applied to a character of $\mathrm{GF}(p^g)$ of order rs in the estimation of $\theta(u,p)$. Here $g = (r-1)(s-1)/2$. Proceed as in part (b).
To prove (a), note that $s_p(\frac{p^f-1}{3}) = f(p-1)/3$. By Lemma 4.1.3, $t\theta(u,p) \leq m/3$. As in part (b), the result follows by Theorem 4.1.4. □

Following Mbodj [92], we say that the pair (u, p) falls under the **index 2 case** if u is odd and $\text{ord}_u(p) = \varphi(u)/2$. Note that u can have at most two distinct prime divisors in this case. The corresponding codes $C(p, m, u)$ will be called **index 2 codes**. Index 2 codes are promising candidates for two-weight codes because of the following.

Proposition 4.3.3 *The number of different nonzero weights of a code $C(p, m, u)$ is at most the number of orbits of $x \mapsto x^p$ on $\mathbb{Z}_u^*$.*

Proof The weight of a codeword $c(y)$ only depends on the coset $\langle \omega \rangle y$. This implies the assertion since the Frobenius automorphism $y \to y^p$ of $\mathbb{F}_{p^m}$ is trace-preserving. □

In particular, an index 2 code with u prime has at most three different nonzero weights. Note that eight of the eleven exceptional two-weight codes listed in Section 4.2.2 are index 2 codes. Thus it is desirable to verify Conjecture 4.2.4 for index 2 codes.

Theorem 4.3.4 *Conditionally on GRH, Conjecture* 4.2.4 *is true for all index* 2 *codes.*

Proof Let $C = C(p, m, u)$ be a two-weight index 2 code. If C is a semiprimitive code, then there is nothing to show. Thus assume that p is not semiprimitive modulo u. First suppose 3 divides u and $p \equiv 1 \pmod 3$. If $u = 3^a s^b$, for a prime $s > 3$, then Theorem 4.3.2 a implies

$$3^{a-1}(s-1)s^{b-1} \leq \frac{3\log((u+1)/4)}{\log p}.$$

Hence,

$$\frac{u \log 7}{12} \leq \log \frac{u+1}{4},$$

a contradiction. The case u is a power of 3 is similar and once again there are no admissible values of u by Theorem 4.3.2 a.
Next suppose that $(u, 3) = 1$. We claim that

$$\frac{\pi\sqrt{u}\log p}{3e\log u} \leq \log \frac{u+1}{4}. \tag{4.12}$$

We carry out the proof of (4.12) only for the case where u has two distinct prime divisors s, r. The case where u is a prime power is similar. Write $u = r^a s^b$ where $a, b \geq 1$. As $\text{ord}_u(p) = \varphi(u)/2$, (4.8) or (4.10) holds for the pair (u, p). If (4.8) holds, then

$$\frac{r^{a-1}s^{b-1}(s-1)}{2} \leq \frac{3e\log r \log{(u+1)/4}}{2\pi\sqrt{r}\log p}$$

by Theorem 4.3.2. If (4.10) holds, then

$$\frac{r^{a-1}s^{b-1}}{2} \le \frac{3e\log rs \log(u+1)/4}{2\pi\sqrt{rs}\log p}$$

by Theorem 4.3.2. Each of these implies (4.12).
Note that (4.12) implies $u < 86,909$ if $p > 2$. Since the table in Section 4.3 contains all exceptional codes with $u \le 100,000$, this shows that Theorem 4.3.4 is true for $p > 2$. If $p = 2$, then (4.12) implies $u < 125,383$. A computer search shows that there are no exceptional codes with $p = 2$ in this range. □

4.4 Two-intersection sets and sub-difference sets

We now discuss the consequences of Theorem 4.1.4 for two-intersection sets in finite projective spaces and sub-difference sets of Singer difference sets. For the connections between these three types of objects, see Section 1.2.6.

4.4.1 Two-intersection sets in PG$(m-1, q)$

Proposition 1.2.17 shows that two-weight irreducible cyclic codes are equivalent to subgroups of Singer cycles which are projective two-intersection sets. Thus Theorem 4.1.4 gives a necessary and sufficient condition for a of a subgroup of the Singer group of PG$(m-1, q)$ to be a two-intersection set. This furnishes a proof for the two-intersection property of examples recently found by Dover and Batten [7] in PG$(2, 5^3)$ and PG$(2, 7^3)$. Those two examples appear on our list of exceptional solutions as $u = 19$ and $u = 37$, respectively. The problem of finding two-intersection sets in projective planes has received special attention. Until recently, all known examples of sets of type (h_1, h_2) in a projective planes (except those with $h_i = 1$ or $h_i = q+1$) had the property that $(h_1 - h_2)^2 = q$, the order of the plane. In particular, these planes all had square order. The examples of Batten and Dover are interesting in that they do not share this property. In fact, none of the exceptional two-intersection sets has the property that the square of the difference of the intersection numbers equals the order of the underlying geometry.

4.4.2 Sub-difference sets of Singer difference sets

Proposition 1.2.19 shows that two-weight irreducible cyclic codes and sub-difference sets of Singer difference sets are equivalent objects. In this section, we indentify all sub-difference sets corresponding to the known two-weight irreducible cyclic codes among the known difference sets. We find it remarkable that not less than five different types of difference sets correspond to the eleven exceptional codes, see the table below.
From the proof of Theorem 4.1.4, we have that E is a (u, k, λ)-difference set in $\mathbb{F}_{p^m}^*/\mathbb{F}_p^*$, where k is from (4.2). It follows immediately that the sub-difference

sets corresponding to the semi-primitive codes are trivial. Similarly, it is straightforward to check that the sub-difference sets corresponding to the one-weight irreducible cyclic codes are again Singer difference sets.
The following table lists the sub-difference sets corresponding to the known exceptional codes. Each of the difference sets on the above list except the $(43,21,10)$ Hall difference set is determined up to equivalence by its parameters (u,k,λ) and the condition that it admits p as a multiplier.

u	k	λ	name	p
11	5	2	QR	3
19	9	4	QR	5
35	17	8	Twin	3
37	9	2	4th	7
43	21	10	Hall	11
67	33	16	QR	17
107	53	26	QR	3
133	33	8	Hall Sp.	5
163	81	40	QR	41
323	161	80	Twin	3
499	249	124	QR	5

Here **QR** stands for the quadratic residue difference set modulo u, and **Twin** denotes the twin-prime power difference sets due to Stanton and Sprott [123]. Moreover, **4th** denotes the set of fourth powers modulo u. Finally, **Hall Sp.** is the $(133,33,8)$ sporadic example found by M. Hall [53].
There are two inequivalent $(43,21,10)$ difference sets in $\mathbb{Z}/43\mathbb{Z}$ admitting the multiplier 11, the quadratic residues and the so-called Hall difference set. Note that 19 is a primitive element and $19^6 \equiv 11 \pmod{43}$. Let

$$C_i = \{19^{i+6j} \mid j = 0,\ldots,6\},$$

for $i = 0,\ldots,5$. The quadradic residues are $QR = C_0 \cup C_2 \cup C_4$ and the Hall difference set is $H = C_0 \cup C_1 \cup C_3$. Pick $y \in C_4$. Consider $\sigma_y \in \mathrm{Gal}(\mathbb{Q}(\xi_{43})/\mathbb{Q})$ defined by $\sigma_y : \xi_{43} \mapsto \xi_{43}^y$. Let $L = \mathbb{F}_{43^{11}}$ and let χ be a character of L^* of order 43. Let D be the Singer difference set of $\mathrm{PG}(6,11)$ and let E be its sub-difference set in $\mathbb{Z}/43\mathbb{Z}$. By a result from [132], we have $\chi(D) = 11G(\chi)$ where $G(\chi)$ is the Gauss sum over L corresponding to χ. From Lemma 1.3.7 we get

$$G(\chi) = 11^3\chi(E).$$

Furthermore, using Result 1.4.9 one checks that

$$(G(\chi))^{\sigma_y} \neq (G(\chi)).$$

It follows that the E cannot be equivalent to QR and therefore is equivalent to the Hall difference set.

The following is equivalent to Conjecture 4.2.4.

Conjecture 4.4.1 *Any nontrivial sub-difference set of a Singer difference set is equivalent either to a Singer difference set or to one of the eleven examples in the above table.*

Bibliography

[1] K.T. Arasu, Y.Q. Chen: A difference set in $(\mathbb{Z}/4\mathbb{Z})^3 \times \mathbb{Z}/5\mathbb{Z}$. Submitted.

[2] K.T. Arasu, J.A. Davis, J. Jedwab: A nonexistence result for abelian Menon difference sets using perfect binary arrays. *Combinatorica* **15** (1995), 311-317.

[3] K.T. Arasu, J.A. Davis, J. Jedwab, S.L. Ma, R.L. McFarland: Exponent bounds for a family of abelian difference sets. *In: Groups, Difference Sets, and the Monster.* Eds. K.T. Arasu et al., DeGruyter Verlag, Berlin/New York 1996, 129-143.

[4] K.T. Arasu, J.F. Dillon, D. Jungnickel, A. Pott: The solution of the Waterloo problem. *J. Comb. Theory Ser. A* **71** (1995), 316-331.

[5] K.T. Arasu, S.K. Sehgal: Difference sets in abelian groups of p-rank two. *Des. Codes Cryptogr.* **5** (1995), 5-12.

[6] K.T. Arasu, Q. Xiang: Multiplier Theorems. *J. Comb. Des.* **3** (1995), 257-267.

[7] L. Batten, J.M. Dover: Some sets of type (m, n) in cubic order planes. *Des. Codes Cryptogr.* **16** (1999), 211-213.

[8] L.D. Baumert: *Cyclic Difference Sets.* Lecture Notes 182, Springer, Berlin/Heidelberg/New York 1971.

[9] L. D. Baumert, R. J. McEliece: Weights of irreducible cyclic codes. *Inform. and Control* **20** (1972), 158-175.

[10] L. D. Baumert, W.H. Mills, R.L. Ward: Uniform cyclotomy. *J. Number Theory* **14** (1982), 67-82.

[11] L. D. Baumert, J. Mykkelveit: *Weight Distribution of Some Irreducible Cyclic Codes.* D.S.N. report **11** (1973), 128-131.

[12] T. Beth, D. Jungnickel, H. Lenz: *Design Theory* (2nd edition). Cambridge University Press, Cambridge 1999.

[13] Z.I. Borevich, I.R. Shafarevich: *Number Theory.* Academic Press, New York/San Francisco/London 1966.

[14] R.C. Bose: An affine analogue of Singer's theorem. *J. Indian Math. Soc.* **6** (1942), 1-15.

[15] R.H. Bruck, H.J. Ryser: The nonexistence of certain finite projective planes. *Can. J. Math.* **1** (1949), 88-93.

[16] A.T. Butson: Generalized Hadamard matrices. *Proc. Amer. Math. Soc.* **13** (1962), 894-898.

[17] A.T. Butson: Relations among generalized Hadamard matrices, relative difference sets, and maximal length linear recurring sequences. *Can. J. Math.* **15** (1963), 42-48.

[18] R. Calderbank, W.M. Kantor: The geometry of two-weight codes. *Bull. London Math. Soc.* **18** (1986), 97-122.

[19] J.W.S. Cassels, A. Froehlich: *Algebraic number theory.* Academic Press, London 1967.

[20] W.K. Chan: Necessary Conditions for Menon Difference Sets. *Des. Codes Cryptogr.* **3** (1993), 147-154.

[21] W.K. Chan, S.L. Ma, M.K. Siu: Non-existence of certain perfect arrays. *Discrete Math.* **125** (1994), 107-113.

[22] Y.Q. Chen: On the existence of abelian Hadamard difference sets and a new family of difference sets. *Finite Fields Appl.* **3** (1997), 234-256.

[23] Y.Q. Chen, D.K. Ray-Chaudhuri, Q. Xiang: Constructions of Partial Difference Sets and Relative Difference Sets Using Galois Rings II. *J. Combin. Theory Ser. A* **76** (1996), 179-196.

[24] G. Cornell: Abhyankar's lemma and the class group. In: *Number Theory.* Ed. M. Nathanson, Lecture Notes in Math. 751, Springer, Berlin/Heidelberg/New York 1979, 82-88.

[25] G. Cornell, M. Rosen: Group-Theoretic Constraints on the Structure of the Class Group. *J. Number Theory* **13** (1981), 1-11.

[26] G. Cornell, L.C. Washington: Class Numbers of Cyclotomic Fields. *J. Number Theory* **21** (1985), 260-274.

[27] R. Craigen: The structure of weighing matrices having large weights. *Des. Codes Cryptogr.* **5** (1995), 199-216.

[28] R. Craigen, H. Kharaghani: Hadamard matrices from weighing matrices via signed groups. *Des. Codes Cryptogr.* **12** (1997), 49-58.

[29] C.W. Curtis, I. Reiner: *Representation Theory of Finite Groups and Associative Algebras.* Wiley, New York/London 1962.

[30] J.A. Davis: Difference sets in abelian 2-groups. *J. Comb. Theory Ser. A* **57** (1991), 262-286.

[31] J.A. Davis: A note on products of relative difference sets. *Des. Codes Cryptogr.* **1** (1991), 117-119.

[32] J.A. Davis: Constructions of relative difference sets in p-groups. *Discrete Math.* **103** (1992), 7-15.

[33] J.A. Davis, J. Jedwab: A unifying construction of difference sets. *J. Comb. Theory Ser. A* **80** (1997), 13-78.

[34] J.A. Davis, J. Jedwab: Nested Hadamard Difference Sets. *J. Statist. Plann. Inference* **62** (1997), 13-20.

[35] J.A. Davis, J. Jedwab, M. Mowbray: New Families of Semi-Regular Relative Difference Sets. *Des. Codes Cryptogr.* **13** (1998), 131-146.

[36] J.A. Davis, S.K. Sehgal: Using the Simplex code to construct relative difference sets in 2-groups. *Des. Codes Cryptogr.* **11** (1997), 267-277.

[37] P. Dembowski: *Finite Geometries.* Springer, Berlin/Heidelberg/New York 1968.

[38] P. Dembowski, F. Piper: Quasiregular collineation groups of finite projective planes. *Math. Zeitschrift* **103** (1967), 53-75.

[39] J.F. Dillon: Variations on a scheme of McFarland for noncyclic difference sets. *J. Combin. Theory Ser. A* **40** (1985), 9-21.

[40] P. Eades, R.M. Hain: On Circulant Weighing Matrices. *Ars Comb.* **2** (1976), 265-284.

[41] S. Eliahou, M. Kervaire: Barker sequences and difference sets. *L'Enseignement Math.* **38** (1992), 345-382.

[42] S. Eliahou, M. Kervaire, B. Saffari: A new restriction on the length of Golay complementary sequences. *J. Comb. Theory Ser. A* **55** (1990), 49-59.

[43] J.E.H. Elliott, A.T. Butson: Relative difference sets. *Illinois J. Math.* **10** (1966), 517-531.

[44] R. Evans: Pure Gauss sums over finite fields. *Mathematika* **28** (1981), 239-248.

[45] M.J. Ganley: On a paper of Dembowski and Ostrom. *Arch. Math.* **27** (1976), 93-98.

[46] A.V. Geramita, J.M. Geramita, J. Seberry: Orthogonal Designs. *J. Lin. Multilin. Algebra* **3** (1975/76), 281-306.

[47] A.V. Geramita, J. Seberry: Orthogonal designs III. Weighing matrices. *Utilitas Math.* **6** (1974), 209-236.

[48] D. Gluck: A note on permutation polynomials and finite geometries. *Discrete Math.* **80** (1990), 97-100.

[49] E.S. Golod, I.R. Shafarevic: On Class Field Towers. *Amer. Math. Soc. Transl. Ser. 2* **48** (1965), 91-102.

[50] M. Gysin, J. Seberry: On the weighing matrices of order $4n$ and weight $4n-2$ and $2n-1$. *Australas. J. Combin.* **12** (1995), 157-174.

[51] J. Hadamard: Résolution d'une question relative aux déterminants. *Bull. Sci. Math.* **2** (1893), 240-246.

[52] M. Hall: Cyclic projective planes. *Duke Math. J.* **14** (1947), 1079-1090.

[53] M. Hall: A survey of difference sets. *Proc. Amer. Math. Soc.* **7** (1956), 975-986.

[54] G.H. Hardy, E.M. Wright: *An Introduction to the Theory of Numbers* (5th edition). Oxford University Press 1979.

[55] H. Hasse: *Über die Klassenzahl Abelscher Zahlkörper.* Springer, Berlin/Heidelberg/New York 1985.

[56] Y. Hiramine: A conjecture on affine planes of prime order. *J. Comb. Theory Ser. A* **52** (1989), 44-50.

[57] J.W.P. Hirschfeld: *Projective Geometries over Finite Fields.* (2nd edition) Oxford University Press 1998.

[58] D.R. Hughes, F. Piper: *Projective Planes.* Springer, Berlin/Heidelberg/New York 1973.

[59] J.E. Iiams: On Difference Sets in Groups of Order $4p^2$. *J. Comb. Theory Ser. A* **72** (1995), 256-276.

[60] J.E. Iiams, R.A. Liebler, K.W. Smith: Difference sets in nilpotent groups with large Frattini quotient: Geometric methods and $(375, 34, 3)$. In: *Groups, Difference Sets and the Monster.* Eds. K.T. Arasu et al., De-Gruyter Verlag, Berlin/New York 1996, 157-168.

[61] K. Ireland, M. Rosen: *A Classical Introduction to Modern Number Theory.* Graduate Texts in Math. 84, Springer, Berlin/Heidelberg/New York 1990.

[62] N. Jacobson: *Basic Algebra I, II* (2nd edition). W. H. Freeman and Company, New York 1985.

[63] D. Jungnickel: On a theorem of Ganley. *Graphs and Comb.* **3** (1987), 141-143.

[64] D. Jungnickel: Difference Sets. In: *Contemporary Design Theory: A Collection of Surveys.* Eds. J.H. Dinitz and D.R. Stinson, Wiley, New York 1992, 241-324.

[65] D. Jungnickel, B. Schmidt: Difference Sets: An Update. In: *Geometry, Combinatorial Designs and Related Structures.* Proceedings of the First Pythagorean Conference. Eds. J.W.P. Hirschfeld et al., Cambridge University Press 1997, 89-112.

[66] D. Jungnickel and B. Schmidt: Difference sets: A second update. Rend. Circ. Mat. Palermo (2) Suppl. **53** (1998), 89-118.

[67] C. Koukouvinos, J. Seberry: Weighing matrices and their applications. *J. Statist. Plann. Inference* **62** (1997), 91-101.

[68] D. Kubert: The 2-divisibility of the class number of cyclotomic fields and the Stickelberger ideal. *J. Reine Angew. Math.* **369** (1986), 192-218.

[69] C.W.H. Lam, L. Thiel, S. Swiercz: The non-existence of finite projective planes of order 10. *Can. J. Math.* **41** (1989), 1117-1123.

[70] E.S. Lander: *Symmetric Designs: An Algebraic Approach.* London Math. Soc. Lect. Notes 75, Cambridge University Press 1983.

[71] P. Langevin: A new class of two weight codes. In: *Finite fields and applications, Glasgow 1995.* London Math. Soc. Lecture Note Ser. 233, Cambridge Univ. Press 1996, 181-187.

[72] P. Langevin: Calculus de certaines sommes de Gauss. *J. Number Theory* **32** (1997), 59-64.

[73] W. de Launey: On the nonexistence of generalized weighing matrices. *Ars Comb.* **21** (1984), 117-132.

[74] W. de Launey, P. Vijay Kumar: On circulant generalized Hadamard matrices of prime power order. Unpublished manuscript (1985).

[75] K.H. Leung, S.L. Ma: Constructions of partial difference sets and relative difference sets on p-groups. *Bull. London Math. Soc.* **22** (1990), 533-539.

[76] R. Lidl, H. Niederreiter: *Introduction to finite fields and their applications.* Cambridge University Press 1994.

[77] R.A. Liebler: The inversion formula. *J. Comb. Math. Comb. Comput.* **13**, (1993), 143-160.

[78] R.A. Liebler, K.W. Smith: On difference sets in certain 2-groups. In: *Coding Theory, Design Theory, Group Theory: Proceedings of the Marshall Hall Conference.* Wiley, New York 1993, 195-212.

[79] C. Lin, W.D. Wallis: Barker sequences and circulant Hadamard matrices. *J. Comb. Inf. Syst. Sci.* **18** (1993), 19-25.

[80] J.H. van Lint: *Introduction to Coding Theory* (3rd edition). Graduate Texts in Mathematics 86, Springer, Berlin/Heidelberg/New York 1999.

[81] S. Louboutin: Minorations (sous l'hypothèse de Riemann généralisée) des nombres de classes des corps quadratiques imaginaires. Application. *C.R. Acad. Sci. Paris Ser. I Math.* **310** (1990), 795-800.

[82] S.L. Ma: *Polynomial addition sets.* Ph.D. thesis, University of Hong Kong, 1985.

[83] S.L. Ma: Planar Functions, Relative Difference Sets and Character Theory. *J. Algebra* **185** (1996), 342-356.

[84] S.L. Ma, A. Pott: Relative difference sets, planar functions and generalized Hadamard matrices. *J. Algebra* **175** (1995), 505-525.

[85] S.L. Ma, B. Schmidt: On (p^a, p, p^a, p^{a-1})-relative difference sets. *Des. Codes Cryptogr.* **6** (1995), 57-72.

[86] S.L. Ma, B. Schmidt: A Sharp Exponent Bound for McFarland Difference Sets with $p = 2$. *J. Combin. Theory Ser. A* **80** (1997), 347-352.

[87] S.L. Ma, B. Schmidt: Relative (p^a, p^b, p^a, p^{b-a})-difference sets: A Unified Exponent Bound and a Local Ring Construction. *Finite Fields Appl.*, to appear.

[88] B.R. MacDonald: *Finite rings with identity.* Marcel Dekker, New York 1974.

[89] H.B. Mann: *Introduction to Algebraic Number Theory.* Ohio State University Press, Columbus 1955.

[90] H.B. Mann: *Addition Theorems.* Wiley, New York 1965.

[91] J. Masley: *On the Class Number of Cyclotomic Fields.* Ph.D. thesis, Princeton University, 1972.

[92] O.D. Mbodj: Quadratic Gauss Sums. *Finite Fields Appl.* **4** (1998), 347-361.

[93] R. J. McEliece: Irreducible cyclic codes and Gauss sums. In: *Combinatorics.* Proc. NATO Advanced Study Inst., Breukelen 1974. Math. Centre Tracts 55, Math. Centrum, Amsterdam 1974, 179-196.

[94] R.L. McFarland: *On multipliers of abelian difference sets.* Ph.D. thesis, Ohio State University, 1970.

[95] R.L. McFarland: A family of difference sets in non-cyclicgroups. *J. Comb. Theory Ser. A* **15** (1973), 1-10.

[96] R.L. McFarland: Difference sets in abelian groups of order $4p^2$. *Mitt. Math. Sem. Giessen* **192** (1989), 1-70.

[97] R.L. McFarland: Sub-difference sets of Hadamard difference sets. *J. Comb. Theory Ser. A* **54** (1990), 112-122.

[98] R.L. McFarland: Necessary conditions for Hadamard difference sets. In: *Coding theory and design theory, Part II.* IMA Vol. Math. Appl. 21, Springer, Berlin/Heidelberg/New York 1990, 257-272.

[99] T. Metsänkylä: Über den ersten Faktor der Klassenzahl des Kreiskörpers. *Ann. Acad. Sci. Fenn. Ser. AI* **416** (1967).

[100] T. Metsänkylä: Über die Teilbarkeit des ersten Faktors der Klassenzahl des Kreiskörpers. *Ann. Univ. Turku. Ser. AI* **124** (1968).

[101] R.C. Mullin: A note on balanced weighing matrices. In: *Combinatorial Mathematics III*. Springer, Berlin/Heidelberg/New York 1985, 28-41.

[102] R.C. Mullin, R.G. Stanton: On the non-existence of a class of circulant balanced weighing matrices. *SIAM J. Appl. Math.* **30** (1976), 98-102.

[103] H. Ohmori: Classification of weighing matrices of order 12 and weight 9. *Discrete Math.* **116** (1993), 55-78.

[104] A. Pott: On the structure of abelian groups admitting divisible difference sets. *J. Comb. Theory Ser. A* **65** (1994), 202-213.

[105] A. Pott: *Finite geometry and character theory.* Lecture Notes 1601, Springer, Berlin/Heidelberg/New York 1995.

[106] A. Pott: A survey on relative difference sets. In: *Groups, Difference Sets and the Monster.* Eds. K.T. Arasu et al., DeGruyter Verlag, Berlin/New York 1996, 195-233.

[107] D.K. Ray-Chaudhuri, Q. Xiang: Constructions of Partial Difference Sets and Relative Difference Sets Using Galois Rings. *Des. Codes Cryptogr.* **8** (1996), 215-227.

[108] D.K. Ray-Chaudhuri, Q. Xiang: New Necessary Conditions for Abelian Hadamard Difference Sets. *J. Statist. Plann. Inference* **62** (1997), 69-79.

[109] P. Ribenboim: *Algebraic Numbers.* Wiley, New York 1972.

[110] P. Ribenboim: *13 Lectures on Fermat's Last Theorem.* Springer, Berlin/Heidelberg/New York 1979.

[111] L. Ronayi, T. Szönyi: Planar functions over finite fields. *Combinatorica* **9** (1989), 315-320.

[112] H.J. Ryser: *Combinatorial Mathematics.* Wiley, New York 1963.

[113] B. Schmidt: *Differenzmengen und relative Differenzmengen.* Dissertation. Verlag Dr. Wißner, Augsburg 1995.

[114] B. Schmidt: On (p^a, p^b, p^a, p^{a-b})-relative difference sets. *J. Alg. Combin.* **6** (1997), 279-297.

[115] B. Schmidt: Nonexistence Results on Chen and Davis-Jedwab Difference Sets. *J. Algebra* **202** (1998), 404-413.

[116] B. Schmidt: Cyclotomic Integers of Prescribed Absolute Value and the Class Group. *J. Number Theory* **72** (1998), 269-281.

[117] B. Schmidt: Cyclotomic Integers and Finite Geometry. *J. Am. Math. Soc.* **12** (1999), 929-952.

[118] B. Schmidt, C. White: All Two-Weight Irreducible Cyclic Codes?. Submitted.

[119] R. Schoof: Infinite class field towers of quadratic fields. *J. Reine Angew. Math.* **372** (1986), 209-220.

[120] J. Seberry, A.L. Whiteman: Some results on weighing matrices. *Bull. Austral. Math. Soc.* **12** (1975), 433-447.

[121] J. Singer: A theorem in finite projective geometry and some applications to number theory. *Trans. Amer. Math. Soc.* **43** (1938), 377-385.

[122] H.Y. Song, S.W. Golomb: On the existence of cyclic Hadamard difference sets. *IEEE Trans. Inform. Th.* **40** (1994), 1266-1268.

[123] R.G. Stanton, D.A. Sprott: A family of difference sets. *Canadian J. Math.* **10** (1958), 73-77.

[124] L. Stickelberger: Über eine Verallgemeinerung der Kreistheilung. *Math. Annalen* **37** (1890), 321-367.

[125] T. Storer: *Cyclotomy and difference sets.* Lectures in Advanced Mathematics 2, Markham Publishing Co., Chicago 1967.

[126] J. Storer, R. Turyn: On binary sequences. *Proc. Amer. Math. Soc.* **12** (1961), 394-399.

[127] R.J. Turyn: Character sums and difference sets. *Pacific J. Math.* **15** (1965), 319-346.

[128] R.J. Turyn: Sequences with small correlation. In: *Error Correcting Codes.* Ed. H.B. Mann, Wiley, New York 1969, 195-228.

[129] R.J. Turyn: Private communication.

[130] L.C. Washington: *Introduction to Cyclotomic Fields.* Graduate Texts in Math. 83, Springer, Berlin/Heidelberg/New York 1997.

[131] R.M. Wilson: An existence theory for pairwise balanced designs, III. Proof of the existence conjectures. *J. Comb. Theory Ser. A* **18** (1975), 71-79.

[132] K. Yamamoto: On congruences arising from relative Gauss sums. In: *Number Theory and Combinatorics.* World Scientific Publ. 1985, 423-446.

[133] K. Yamamoto: Decomposition fields of difference sets. *Pacific J. Math.* **13** (1963), 337-383.

Index

Lecture Notes in Mathematics

For information about Vols. 1–1619
please contact your bookseller or Springer-Verlag

Vol. 1620: R. Donagi, B. Dubrovin, E. Frenkel, E. Previato, Integrable Systems and Quantum Groups. Montecatini Terme, 1993. Editors:M. Francaviglia, S. Greco. VIII, 488 pages. 1996.

Vol. 1621: H. Bass, M. V. Otero-Espinar, D. N. Rockmore, C. P. L. Tresser, Cyclic Renormalization and Automorphism Groups of Rooted Trees. XXI, 136 pages. 1996.

Vol. 1622: E. D. Farjoun, Cellular Spaces, Null Spaces and Homotopy Localization. XIV, 199 pages. 1996.

Vol. 1623: H.P. Yap, Total Colourings of Graphs. VIII, 131 pages. 1996.

Vol. 1624: V. Brýnzanescu, Holomorphic Vector Bundles over Compact Complex Surfaces. X, 170 pages. 1996.

Vol. 1625: S. Lang, Topics in Cohomology of Groups. VII, 226 pages. 1996.

Vol. 1626: J. Azéma, M. Emery, M. Yor (Eds.), Séminaire de Probabilités XXX. VIII, 382 pages. 1996.

Vol. 1627: C. Graham, Th. G. Kurtz, S. Méléard, Ph. E. Protter, M. Pulvirenti, D. Talay, Probabilistic Models for Nonlinear Partial Differential Equations. Montecatini Terme, 1995. Editors: D. Talay, L. Tubaro. X, 301 pages. 1996.

Vol. 1628: P.-H. Zieschang, An Algebraic Approach to Association Schemes. XII, 189 pages. 1996.

Vol. 1629: J. D. Moore, Lectures on Seiberg-Witten Invariants. VII, 105 pages. 1996.

Vol. 1630: D. Neuenschwander, Probabilities on the Heisenberg Group: Limit Theorems and Brownian Motion. VIII, 139 pages. 1996.

Vol. 1631: K. Nishioka, Mahler Functions and Transcendence. VIII, 185 pages. 1996.

Vol. 1632: A. Kushkuley, Z. Balanov, Geometric Methods in Degree Theory for Equivariant Maps. VII, 136 pages. 1996.

Vol. 1633: H. Aikawa, M. Essén, Potential Theory – Selected Topics. IX, 200 pages. 1996.

Vol. 1634: J. Xu, Flat Covers of Modules. IX, 161 pages. 1996.

Vol. 1635: E. Hebey, Sobolev Spaces on Riemannian Manifolds. X, 116 pages. 1996.

Vol. 1636: M. A. Marshall, Spaces of Orderings and Abstract Real Spectra. VI, 190 pages. 1996.

Vol. 1637: B. Hunt, The Geometry of some special Arithmetic Quotients. XIII, 332 pages. 1996.

Vol. 1638: P. Vanhaecke, Integrable Systems in the realm of Algebraic Geometry. VIII, 218 pages. 1996.

Vol. 1639: K. Dekimpe, Almost-Bieberbach Groups: Affine and Polynomial Structures. X, 259 pages. 1996.

Vol. 1640: G. Boillat, C. M. Dafermos, P. D. Lax, T. P. Liu, Recent Mathematical Methods in Nonlinear Wave Propagation. Montecatini Terme, 1994. Editor: T. Ruggeri. VII, 142 pages. 1996.

Vol. 1641: P. Abramenko, Twin Buildings and Applications to S-Arithmetic Groups. IX, 123 pages. 1996.

Vol. 1642: M. Puschnigg, Asymptotic Cyclic Cohomology. XXII, 138 pages. 1996.

Vol. 1643: J. Richter-Gebert, Realization Spaces of Polytopes. XI, 187 pages. 1996.

Vol. 1644: A. Adler, S. Ramanan, Moduli of Abelian Varieties. VI, 196 pages. 1996.

Vol. 1645: H. W. Broer, G. B. Huitema, M. B. Sevryuk, Quasi-Periodic Motions in Families of Dynamical Systems. XI, 195 pages. 1996.

Vol. 1646: J.-P. Demailly, T. Peternell, G. Tian, A. N. Tyurin, Transcendental Methods in Algebraic Geometry. Cetraro, 1994. Editors: F. Catanese, C. Ciliberto. VII, 257 pages. 1996.

Vol. 1647: D. Dias, P. Le Barz, Configuration Spaces over Hilbert Schemes and Applications. VII. 143 pages. 1996.

Vol. 1648: R. Dobrushin, P. Groeneboom, M. Ledoux, Lectures on Probability Theory and Statistics. Editor: P. Bernard. VIII, 300 pages. 1996.

Vol. 1649: S. Kumar, G. Laumon, U. Stuhler, Vector Bundles on Curves – New Directions. Cetraro, 1995. Editor: M. S. Narasimhan. VII, 193 pages. 1997.

Vol. 1650: J. Wildeshaus, Realizations of Polylogarithms. XI, 343 pages. 1997.

Vol. 1651: M. Drmota, R. F. Tichy, Sequences, Discrepancies and Applications. XIII, 503 pages. 1997.

Vol. 1652: S. Todorcevic, Topics in Topology. VIII, 153 pages. 1997.

Vol. 1653: R. Benedetti, C. Petronio, Branched Standard Spines of 3-manifolds. VIII, 132 pages. 1997.

Vol. 1654: R. W. Ghrist, P. J. Holmes, M. C. Sullivan, Knots and Links in Three-Dimensional Flows. X, 208 pages. 1997.

Vol. 1655: J. Azéma, M. Emery, M. Yor (Eds.), Séminaire de Probabilités XXXI. VIII, 329 pages. 1997.

Vol. 1656: B. Biais, T. Björk, J. Cvitanic, N. El Karoui, E. Jouini, J. C. Rochet, Financial Mathematics. Bressanone, 1996. Editor: W. J. Runggaldier. VII, 316 pages. 1997.

Vol. 1657: H. Reimann, The semi-simple zeta function of quaternionic Shimura varieties. IX, 143 pages. 1997.

Vol. 1658: A. Pumarino, J. A. Rodrýguez, Coexistence and Persistence of Strange Attractors. VIII, 195 pages. 1997.

Vol. 1659: V, Kozlov, V. Maz'ya, Theory of a Higher-Order Sturm-Liouville Equation. XI, 140 pages. 1997.

Vol. 1660: M. Bardi, M. G. Crandall, L. C. Evans, H. M. Soner, P. E. Souganidis, Viscosity Solutions and Applications. Montecatini Terme, 1995. Editors: I. Capuzzo Dolcetta, P. L. Lions. IX, 259 pages. 1997.

Vol. 1661: A. Tralle, J. Oprea, Symplectic Manifolds with no Kähler Structure. VIII, 207 pages. 1997.

Vol. 1662: J. W. Rutter, Spaces of Homotopy Self-Equivalences – A Survey. IX, 170 pages. 1997.

Vol. 1663: Y. E. Karpeshina; Perturbation Theory for the Schrödinger Operator with a Periodic Potential. VII, 352 pages. 1997.

Vol. 1664: M. Väth, Ideal Spaces. V, 146 pages. 1997.

Vol. 1665: E. Giné, G. R. Grimmett, L. Saloff-Coste, Lectures on Probability Theory and Statistics 1996. Editor: P. Bernard. X, 424 pages, 1997.

Vol. 1666: M. van der Put, M. F. Singer, Galois Theory of Difference Equations. VII, 179 pages. 1997.

Vol. 1667: J. M. F. Castillo, M. González, Three-space Problems in Banach Space Theory. XII, 267 pages. 1997.

Vol. 1668: D. B. Dix, Large-Time Behavior of Solutions of Linear Dispersive Equations. XIV, 203 pages. 1997.

Vol. 1669: U. Kaiser, Link Theory in Manifolds. XIV, 167 pages. 1997.

Vol. 1670: J. W. Neuberger, Sobolev Gradients and Differential Equations. VIII, 150 pages. 1997.

Vol. 1671: S. Bouc, Green Functors and G-sets. VII, 342 pages. 1997.

Vol. 1672: S. Mandal, Projective Modules and Complete Intersections. VIII, 114 pages. 1997.

Vol. 1673: F. D. Grosshans, Algebraic Homogeneous Spaces and Invariant Theory. VI, 148 pages. 1997.

Vol. 1674: G. Klaas, C. R. Leedham-Green, W. Plesken, Linear Pro-p-Groups of Finite Width. VIII, 115 pages. 1997.

Vol. 1675: J. E. Yukich, Probability Theory of Classical Euclidean Optimization Problems. X, 152 pages. 1998.

Vol. 1676: P. Cembranos, J. Mendoza, Banach Spaces of Vector-Valued Functions. VIII, 118 pages. 1997.

Vol. 1677: N. Proskurin, Cubic Metaplectic Forms and Theta Functions. VIII, 196 pages. 1998.

Vol. 1678: O. Krupková, The Geometry of Ordinary Variational Equations. X, 251 pages. 1997.

Vol. 1679: K.-G. Grosse-Erdmann, The Blocking Technique. Weighted Mean Operators and Hardy's Inequality. IX, 114 pages. 1998.

Vol. 1680: K.-Z. Li, F. Oort, Moduli of Supersingular Abelian Varieties. V, 116 pages. 1998.

Vol. 1681: G. J. Wirsching, The Dynamical System Generated by the 3n+1 Function. VII, 158 pages. 1998.

Vol. 1682: H.-D. Alber, Materials with Memory. X, 166 pages. 1998.

Vol. 1683: A. Pomp, The Boundary-Domain Integral Method for Elliptic Systems. XVI, 163 pages. 1998.

Vol. 1684: C. A. Berenstein, P. F. Ebenfelt, S. G. Gindikin, S. Helgason, A. E. Tumanov, Integral Geometry, Radon Transforms and Complex Analysis. Firenze, 1996. Editors: E. Casadio Tarabusi, M. A. Picardello, G. Zampieri. VII, 160 pages. 1998.

Vol. 1685: S. König, A. Zimmermann, Derived Equivalences for Group Rings. X, 146 pages. 1998.

Vol. 1686: J. Azéma, M. Émery, M. Ledoux, M. Yor (Eds.), Séminaire de Probabilités XXXII. VI, 440 pages. 1998.

Vol. 1687: F. Bornemann, Homogenization in Time of Singularly Perturbed Mechanical Systems. XII, 156 pages. 1998.

Vol. 1688: S. Assing, W. Schmidt, Continuous Strong Markov Processes in Dimension One. XII, 137 page. 1998.

Vol. 1689: W. Fulton, P. Pragacz, Schubert Varieties and Degeneracy Loci. XI, 148 pages. 1998.

Vol. 1690: M. T. Barlow, D. Nualart, Lectures on Probability Theory and Statistics. Editor: P. Bernard. VIII, 237 pages. 1998.

Vol. 1691: R. Bezrukavnikov, M. Finkelberg, V. Schechtman, Factorizable Sheaves and Quantum Groups. X, 282 pages. 1998.

Vol. 1692: T. M. W. Eyre, Quantum Stochastic Calculus and Representations of Lie Superalgebras. IX, 138 pages. 1998.

Vol. 1694: A. Braides, Approximation of Free-Discontinuity Problems. XI, 149 pages. 1998.

Vol. 1695: D. J. Hartfiel, Markov Set-Chains. VIII, 131 pages. 1998.

Vol. 1696: E. Bouscaren (Ed.): Model Theory and Algebraic Geometry. XV, 211 pages. 1998.

Vol. 1697: B. Cockburn, C. Johnson, C.-W. Shu, E. Tadmor, Advanced Numerical Approximation of Nonlinear Hyperbolic Equations. Cetraro, Italy, 1997. Editor: A. Quarteroni. VII, 390 pages. 1998.

Vol. 1698: M. Bhattacharjee, D. Macpherson, R. G. Möller, P. Neumann, Notes on Infinite Permutation Groups. XI, 202 pages. 1998.

Vol. 1699: A. Inoue,Tomita-Takesaki Theory in Algebras of Unbounded Operators. VIII, 241 pages. 1998.

Vol. 1700: W. A. Woyczyński, Burgers-KPZ Turbulence, XI, 318 pages. 1998.

Vol. 1701: Ti-Jun Xiao, J. Liang, The Cauchy Problem of Higher Order Abstract Differential Equations, XII, 302 pages. 1998.

Vol. 1702: J. Ma, J. Yong, Forward-Backward Stochastic Differential Equations and Their Applications. XIII, 270 pages. 1999.

Vol. 1703: R. M. Dudley, R. Norvaiša, Differentiability of Six Operators on Nonsmooth Functions and p-Variation. VIII, 272 pages. 1999.

Vol. 1704: H. Tamanoi, Elliptic Genera and Vertex Operator Super-Algebras. VI, 390 pages. 1999.

Vol. 1705: I. Nikolaev, E. Zhuzhoma, Flows in 2-dimensional Manifolds. XIX, 294 pages. 1999.

Vol. 1706: S. Yu. Pilyugin, Shadowing in Dynamical Systems. XVII, 271 pages. 1999.

Vol. 1707: R. Pytlak, Numerical Methods for Optimal Control Problems with State Constraints. XV, 215 pages. 1999.

Vol. 1708: K. Zuo, Representations of Fundamental Groups of Algebraic Varieties. VII, 139 pages. 1999.

Vol. 1709: J. Azéma, M. Émery, M. Ledoux, M. Yor (Eds), Séminaire de Probabilités XXXIII. VIII, 418 pages. 1999.

Vol. 1710: M. Koecher, The Minnesota Notes on Jordan Algebras and Their Applications. IX, 173 pages. 1999.

Vol. 1711: W. Ricker, Operator Algebras Generated by Commuting Projections: A Vector Measure Approach. XVII, 159 pages. 1999.

Vol. 1712: N. Schwartz, J. J. Madden, Semi-algebraic Function Rings and Reflectors of Partially Ordered Rings. XI, 279 pages. 1999.

Vol. 1713: F. Bethuel, G. Huisken, S. Müller, K. Steffen, Calculus of Variations and Geometric Evolution Problems. Cetraro, 1996. Editors: S. Hildebrandt, M. Struwe. VII, 293 pages. 1999.

Vol. 1714: O. Diekmann, R. Durrett, K. P. Hadeler, P. K. Maini, H. L. Smith, Mathematics Inspired by Biology. Martina Franca, 1997. Editors: V. Capasso, O. Diekmann. VII, 268 pages. 1999.

Vol. 1715: N. V. Krylov, M. Röckner, J. Zabczyk, Stochastic PDE's and Kolmogorov Equations in Infinite Dimensions. Cetraro, 1998. Editor: G. Da Prato. VIII, 239 pages. 1999.

Vol. 1716: J. Coates, R. Greenberg, K. A. Ribet, K. Rubin, Arithmetic Theory of Elliptic Curves. Cetraro, 1997. Editor: C. Viola. VIII, 260 pages. 1999.
Vol. 1717: J. Bertoin, F. Martinelli, Y. Peres, Lectures on Probability Theory and Statistics. Saint-Flour, 1997. Editor: P. Bernard. IX, 291 pages. 1999.
Vol. 1718: A. Eberle, Uniqueness and Non-Uniqueness of Semigroups Generated by Singular Diffusion Operators. VIII, 262 pages. 1999.
Vol. 1719: K. R. Meyer, Periodic Solutions of the N-Body Problem. IX, 144 pages. 1999.
Vol. 1720: D. Elworthy, Y. Le Jan, X-M. Li, On the Geometry of Diffusion Operators and Stochastic Flows. IV, 118 pages. 1999.
Vol. 1721: A. Iarrobino, V. Kanev, Power Sums, Gorenstein Algebras, and Determinantal Loci. XXVII, 345 pages. 1999.
Vol. 1722: R. McCutcheon, Elemental Methods in Ergodic Ramsey Theory. VI, 160 pages. 1999.
Vol. 1723: J. P. Croisille, C. Lebeau, Diffraction by an Immersed Elastic Wedge. VI, 134 pages. 1999.
Vol. 1724: V. N. Kolokoltsov, Semiclassical Analysis for Diffusions and Stochastic Processes. VIII, 347 pages. 2000.
Vol. 1725: D. A. Wolf-Gladrow, Lattice-Gas Cellular Automata and Lattice Boltzmann Models. IX, 308 pages. 2000.
Vol. 1726: V. Marić, Regular Variation and Differential Equations. X, 127 pages. 2000.
Vol. 1727: P. Kravanja M. Van Barel, Computing the Zeros of Analytic Functions. VII, 111 pages. 2000.
Vol. 1728: K. Gatermann Computer Algebra Methods for Equivariant Dynamical Systems. XV, 153 pages. 2000.
Vol. 1729: J. Azéma, M. Émery, M. Ledoux, M. Yor Séminaire de Probabilités XXXIV. VI, 431 pages. 2000.
Vol. 1730: S. Graf, H. Luschgy, Foundations of Quantization for Probability Distributions. X, 230 pages. 2000.
Vol. 1731: T. Hsu, Quilts: Central Extensions, Braid Actions, and Finite Groups. XII, 185 pages. 2000.
Vol. 1732: K. Keller, Invariant Factors, Julia Equivalences and the (Abstract) Mandelbrot Set. X, 206 pages. 2000.
Vol. 1733: K. Ritter, Average-Case Analysis of Numerical Problems. IX, 254 pages. 2000.
Vol. 1734: M. Espedal, A. Fasano, A. Mikelić, Filtration in Porous Media and Industrial Applications. Cetraro 1998. Editor: A. Fasano. 2000.
Vol. 1735: D. Yafaev, Scattering Theory: Some Old and New Problems. XVI, 169 pages. 2000.
Vol. 1736: B. O. Turesson, Nonlinear Potential Theory and Weighted Sobolev Spaces. XIV, 173 pages. 2000.
Vol. 1737: S. Wakabayashi, Classical Microlocal Analysis in the Space of Hyperfunctions. VIII, 367 pages. 2000.
Vol. 1738: M. Émery, A. Nemirovski, D. Voiculescu, Lectures on Probability Theory and Statistics. XI, 356 pages. 2000.
Vol. 1739: R. Burkard, P. Deuflhard, A. Jameson, J.-L. Lions, G. Strang, Computational Mathematics Driven by Industrial Problems. Martina Franca, 1999. Editors: V. Capasso, H. Engl, J. Periaux. VII, 418 pages. 2000.
Vol. 1740: B. Kawohl, O. Pironneau, L. Tartar, J.-P. Zolesio, Optimal Shape Design. Tróia, Portugal 1999. Editors: A. Cellina, A. Ornelas. IX, 388 pages. 2000.
Vol. 1741: E. Lombardi, Oscillatory Integrals and Phenomena Beyond all Algebraic Orders. XV, 413 pages. 2000.
Vol. 1742: A. Unterberger, Quantization and Non-holomorphic Modular Forms. VIII, 253 pages. 2000.
Vol. 1743: L. Habermann, Riemannian Metrics of Constant Mass and Moduli Spaces of Conformal Structures. XII, 116 pages. 2000.
Vol. 1744: M. Kunze, Non-Smooth Dynamical Systems. X, 228 pages. 2000.
Vol. 1745: V. D. Milman, G. Schechtman, Geometric Aspects of Functional Analysis. VIII, 289 pages. 2000.
Vol. 1746: A. Degtyarev, I. Itenberg, V. Kharlamov, Real Enriques Surfaces. XVI, 259 pages. 2000.
Vol. 1747: L. W. Christensen, Gorenstein Dimensions. VIII, 204 pages. 2000.
Vol. 1748: M. Ruzicka, Electrorheological Fluids: Modeling and Mathematical Theory. XV, 176 pages. 2001.
Vol. 1749: M. Fuchs, G. Seregin, Variational Methods for Problems from Plasticity Theory and for Generalized Newtonian Fluids. VI, 269 pages. 2001.
Vol. 1750: B. Conrad, Grothendieck Duality and Base Change. X, 296 pages. 2001.
Vol. 1751: N. J. Cutland, Loeb Measures in Practice: Recent Advances. XI, 111 pages. 2001.
Vol. 1752: Y. V. Nesterenko, P. Philippon, Introduction to Algebraic Independence Theory. XIII, 256 pages. 2001.
Vol. 1753: A. I. Bobenko, U. Eitner, Painlevé Equations in the Differential Geometry of Surfaces. VI, 120 pages. 2001.
Vol. 1754: W. Bertram, The Geometry of Jordan and Lie Structures. XVI, 269 pages. 2001.
Vol. 1755: J. Azéma, M. Émery, M. Ledoux, M. Yor, Séminaire de Probabilités XXXV. VI, 427 pages. 2001.
Vol. 1756: P. E. Zhidkov, Korteweg de Vries and Nonlinear Schrödinger Equations: Qualitative Theory. VII, 147 pages. 2001.
Vol. 1757: R. R. Phelps, Lectures on Choquet's Theorem. VII, 124 pages. 2001.
Vol. 1758: N. Monod, Continuous Bounded Cohomology of Locally Compact Groups. X, 214 pages. 2001.
Vol. 1759: Y. Abe, K. Kopfermann, Toroidal Groups. VIII, 133 pages. 2001.
Vol. 1760: D. Filipović, Consistency Problems for Heath-Jarrow-Morton Interest Rate Models. VIII, 134 pages. 2001.
Vol. 1761: C. Adelmann, The Decomposition of Primes in Torsion Point Fields. VI, 142 pages. 2001.
Vol. 1762: S. Cerrai, Second Order PDE's in Finite and Infinite Dimension. IX, 330 pages. 2001.
Vol. 1763: J.-L. Loday, A. Frabetti, F. Chapoton, F. Goichot, Dialgebras and Related Operads. IV, 132 pages. 2001.
Vol. 1764: A. Cannas da Silva, Lectures on Symplectic Geometry. XII, 217 pages. 2001.
Vol. 1765: T. Kerler, V. V. Lyubashenko, Non-Semisimple Topological Quantum Field Theories for 3-Manifolds with Corners. VI, 379 pages. 2001.
Vol. 1766: H. Hennion, L. Hervé, Limit Theorems for Markov Chains and Stochastic Properties of Dynamical Systems by Quasi-Compactness. VIII, 145 pages. 2001.
Vol. 1767: J. Xiao, Holomorphic Q Classes. VIII, 112 pages. 2001.
Vol. 1768: M.J. Pflaum, Analytic and Geometric Study of Stratified Spaces. VIII, 230 pages. 2001.
Vol. 1769: M. Alberich-Carramiñana, Geometry of the Plane Cremona Maps. XVI, 257 pages. 2002.

Vol. 1770: H. Gluesing-Luerssen, Linear Delay-Differential Systems with Commensurate Delays: An Algebraic Approach. VIII, 176 pages. 2002.

Vol. 1771: M. Émery, M. Yor, Séminaire de Probabilités 1967-1980. A Selection in Martingale Theory. IX, 553 pages. 2002.

Vol. 1772: F. Burstall, D. Ferus, K. Leschke, F. Pedit, U. Pinkall, Conformal Geometry of Surfaces in S^4. VII, 89 pages. 2002.

Vol. 1773: Z. Arad, M. Muzychuk, Standard Integral Table Algebras Generated by a Non-real Element of Small Degree. X, 126 pages. 2002.

Vol. 1774: V. Runde, Lectures on Amenability. XIV, 296 pages. 2002.

Vol. 1775: W. H. Meeks, A. Ros, H. Rosenberg, The Global Theory of Minimal Surfaces in Flat Spaces. Martina Franca 1999. Editor: G. P. Pirola. X, 117 pages. 2002.

Vol. 1776: K. Behrend, C. Gomez, V. Tarasov, G. Tian, Quantum Comohology. Cetraro 1997. Editors: P. de Bartolomeis, B. Dubrovin, C. Reina. VIII, 319 pages. 2002.

Vol. 1777: E. García-Río, D. N. Kupeli, R. Vázquez-Lorenzo, Osserman Manifolds in Semi-Riemannian Geometry. XII, 166 pages. 2002.

Vol. 1778: H. Kiechle, Theory of K-Loops. X, 186 pages. 2002.

Vol. 1779: I. Chueshov, Monotone Random Systems. VIII, 234 pages. 2002.

Vol. 1780: J. H. Bruinier, Borcherds Products on O(2,1) and Chern Classes of Heegner Divisors. VIII, 152 pages. 2002.

Vol. 1781: E. Bolthausen, E. Perkins, A. van der Vaart, Lectures on Probability Theory and Statistics. Ecole d' Eté de Probabilités de Saint-Flour XXIX-1999. Editor: P. Bernard. VIII, 466 pages. 2002.

Vol. 1782: C.-H. Chu, A. T.-M. Lau, Harmonic Functions on Groups and Fourier Algebras. VII, 100 pages. 2002.

Vol. 1783: L. Grüne, Asymptotic Behavior of Dynamical and Control Systems under Perturbation and Discretization. IX, 231 pages. 2002.

Vol. 1784: L.H. Eliasson, S. B. Kuksin, S. Marmi, J.-C. Yoccoz, Dynamical Systems and Small Divisors. Cetraro, Italy 1998. Editors: S. Marmi, J.-C. Yoccoz. VIII, 199 pages. 2002.

Vol. 1785: J. Arias de Reyna, Pointwise Convergence of Fourier Series. XVIII, 175 pages. 2002.

Vol. 1786: S. D. Cutkosky, Monomialization of Morphisms from 3-Folds to Surfaces. V, 235 pages. 2002.

Vol. 1787: S. Caenepeel, G. Militaru, S. Zhu, Frobenius and Separable Functors for Generalized Module Categories and Nonlinear Equations. XIV, 354 pages. 2002.

Vol. 1788: A. Vasil'ev, Moduli of Families of Curves for Conformal and Quasiconformal Mappings.IX, 211 pages. 2002.

Vol. 1789: Y. Sommerhäuser, Yetter-Drinfel'd Hopf algebras over groups of prime order. V, 157 pages. 2002.

Vol. 1790: X. Zhan, Matrix Inequalities. VII, 116 pages. 2002.

Vol. 1791: M. Knebusch, D. Zhang, Manis Valuations and Prüfer Extensions I: A new Chapter in Commutative Algebra. VI, 267 pages. 2002.

Vol. 1792: D. D. Ang, R. Gorenflo, V. K. Le, D. D. Trong, Moment Theory and Some Inverse Problems in Potential Theory and Heat Conduction. VIII, 183 pages. 2002.

Vol. 1793: J. Cortés Monforte, Geometric, Control and Numerical Aspects of Nonholonomic Systems. XV, 219 pages. 2002.

Vol. 1794: N. Pytheas Fogg, Substitution in Dynamics, Arithmetics and Combinatorics. Editors: V. Berthé, S. Ferenczi, C. Mauduit, A. Siegel. XVII, 402 pages. 2002.

Vol. 1795: H. Li, Filtered-Graded Transfer in Using Noncommutative Gröbner Bases. IX, 197 pages. 2002.

Vol. 1796: J.M. Melenk, hp-Finite Element Methods for Singular Perturbations. XIV, 318 pages. 2002.

Vol. 1797: B. Schmidt, Characters and Cyclotomic Fields in Finite Geometry. VIII, 100 pages. 2002.

Vol. 1798: W.M. Oliva, Geometric Mechanics. XI, 270 pages. 2002.

Recent Reprints and New Editions

Vol. 1200: V. D. Milman, G. Schechtman, Asymptotic Theory of Finite Dimensional Normed Spaces. 1986. – Corrected Second Printing. X, 156 pages. 2001.

Vol. 1618: G. Pisier, Similarity Problems and Completely Bounded Maps. 1995 – Second, Expanded Edition VII, 198 pages. 2001.

Vol. 1629: J. D. Moore, Lectures on Seiberg-Witten Invariants. 1997 – Second Edition. VIII, 121 pages. 2001.

Vol. 1638: P. Vanhaecke, Integrable Systems in the realm of Algebraic Geometry. 1996 – Second Edition. X, 256 pages. 2001.

Vol. 1702: J. Ma, J. Yong, Forward-Backward Stochastic Differential Equations and Their Applications. 1999. – Corrected Second Printing. XIII, 270 pages. 2000.